AF805787

Civil Protection, Disaster Management, and Resilience

Volume 4

Editor-in-Chief

Thomas Glade, Department of Geography and Regional Research, University of Vienna, Vienna, Austria

Editorial Board

Peter Billing, Department of Geography, University of Bonn, Bonn, Germany

Aurélia Bony-Dandrieux, Institut des Sciences des Risques, École des Mines d'Alès, Alès, France

Caroline McMullan, DCU Business School, Dublin City University, Dublin, Ireland

Raul Savimaa, Research Centre of Internal Security, Estonian Academy of Security Sciences, Tallinn, Estonia

Lothar Schrott, Department of Geography, University of Bonn, Bonn, Germany

Emanuele Sommario, Dirpolis Institute, Scuola Superiore Sant'Anna, Pisa, Italy

Cees J. van Westen, Faculty of Geo-Information Science and Earth Observation, University of Twente, Enschede, Overijssel, The Netherlands

Assistant Editors

Annika Fröwis, Department of Geography and Regional Research, University of Vienna, Vienna, Austria

Sophia Sternath, Department of Geography and Regional Research, University of Vienna, Vienna, Austria

Landslides, floods, pandemics, geopolitical unrest—disasters are increasing in frequency, severity, complexity, and emerging in unexpected locations. Addressing these escalating challenges demands comprehensive knowledge, innovative concepts, advanced methodologies, and resilience-building strategies.

The new book series, *Civil Protection, Disaster Management, and Resilience*, offers essential insights, practical tools, and proven best practices for navigating today's multifaceted risk landscapes. The series launches with seven foundational volumes aligned with the EU's EUMA postgraduate curriculum, serving as a critical resource for foundational and advanced education and real-world applications. For more details about the EUMA curriculum and its objectives, visit the EUMA homepage (https://postgraduatecenter.univie.ac.at/en/euma).

This book series serves as a foundational resource for professionals across various fields of Disaster Risk Management, as well as for students and researchers in related disciplines.

Looking ahead, the series remains open to new contributions, inviting academics and practitioners to expand its scope with fresh perspectives and emerging ideas. Whether you are shaping policy, advancing academic research, or implementing solutions on the ground, this series equips you to address the underlying causes and key drivers of today's dynamic risks, fostering safer, more resilient societies.

Interested in contributing? Contact the publishing editor or the assisting editor (euma@univie.ac.at) to join this vital and evolving initiative.

Emanuele Sommario · Peter Billing
Editors

International Legal and Institutional Frameworks for Disaster Risk Management

Funded by
the European Union

Springer

Editors
Emanuele Sommario
DIRPOLIS Institute
Scuola Superiore Sant'Anna
Pisa, Italy

Peter Billing
Department of Geography
University of Bonn
Bonn, Germany

ISSN 3059-2704 ISSN 3059-2712 (electronic)
Civil Protection, Disaster Management, and Resilience
ISBN 978-3-032-11805-9 ISBN 978-3-032-11806-6 (eBook)
https://doi.org/10.1007/978-3-032-11806-6

This work was supported by the European Commission, the Scuola Superiore Sant'Anna and the University of Vienna.

© The Editor(s) (if applicable) and The Author(s) 2026. This book is an open access publication.

Open Access This book is licensed under the terms of the Creative Commons Attribution 4.0 International License (http://creativecommons.org/licenses/by/4.0/), which permits use, sharing, adaptation, distribution and reproduction in any medium or format, as long as you give appropriate credit to the original author(s) and the source, provide a link to the Creative Commons license and indicate if changes were made.
The images or other third party material in this book are included in the book's Creative Commons license, unless indicated otherwise in a credit line to the material. If material is not included in the book's Creative Commons license and your intended use is not permitted by statutory regulation or exceeds the permitted use, you will need to obtain permission directly from the copyright holder.
Unless otherwise expressly agreed by the Publisher in writing, the Publisher does not authorize use or reproduction of all or any part of this work in any manner for the purpose of training, fine-tuning, evaluating, or developing artificial intelligence models, technologies, or systems. In accordance with Article 4(3) of the EU Directive 2019/790 on copyright and related rights in the Digital Single Market, the Publisher expressly reserves all rights with respect to this work under the text and data mining exception of the directive.
The use of general descriptive names, registered names, trademarks, service marks, etc. in this publication does not imply, even in the absence of a specific statement, that such names are exempt from the relevant protective laws and regulations and therefore free for general use.
The publisher, the authors and the editors are safe to assume that the advice and information in this book are believed to be true and accurate at the date of publication. Neither the publisher nor the authors or the editors give a warranty, expressed or implied, with respect to the material contained herein or for any errors or omissions that may have been made. The publisher remains neutral with regard to jurisdictional claims in published maps and institutional affiliations.

Cover Image: © 7e RIISC—Sécurité Civile France

Disclaimer: Funded by the European Union. However, the views and opinions expressed are those of the author(s) only and do not necessarily reflect those of the European Union or the European Commission. Neither the European Union nor the granting authority can be held responsible for them.

This Springer imprint is published by the registered company Springer Nature Switzerland AG
The registered company address is: Gewerbestrasse 11, 6330 Cham, Switzerland

If disposing of this product, please recycle the paper.

Foreword by the Commissioner to the EUMA Book Series Volumes

Crises today are more complex and interconnected than ever before. The world is warming at an unprecedented pace, especially in Europe, with global temperatures now approaching 1.5 °C above pre-industrial levels. In this context of growing geopolitical uncertainty, the need for strong preparedness and response systems has never been clearer.

Recent record-breaking wildfires, floods, and other extreme weather events—all linked to climate change—illustrate the mounting impact of a heating planet and the urgency of building resilience not only to natural disasters but also to human-made crises such as conflict. One of the European Commission's top priorities is to strengthen crisis management and civil preparedness across Europe—to move from reacting to crises to being ready for them. This means fostering a genuine culture of preparedness and resilience, one that is woven into every part of our societies—from schools and communities to governments and businesses.

But we cannot do this alone. Collaboration is key—and the academic community has a crucial role to play. By educating students and practitioners in Disaster Risk Management, universities help us raise awareness, build a shared understanding of the risks we face, and inspire the innovation and mindset needed to turn preparedness into a way of life.

This book series marks an important milestone for the EUMA project—Creating a European Higher Education Network for Master's Programmes in Disaster Risk Management—and for the postgraduate master's degree in International Disaster Management and Civil Protection it has developed. Supported by the EU's preparedness capacity-building tools, the EUMA series represents a renewed drive to bring research, innovation, and practical action closer together. It reflects the European Commission's commitment to embedding knowledge-based strategies into preparedness and response and demonstrates the concrete value of EU programmes that strengthen our collective resilience.

The Preparedness Union Strategy is our compass—guiding us towards a new level of collective readiness that is proactive, interconnected, and truly integrated. This book series supports one of its key actions: bringing preparedness into education. By focusing on higher education and developing a robust master's programme, the EUMA project has helped build a holistic approach to preparedness—one that equips future generations with the knowledge and skills needed to manage disaster risks effectively.

I want to warmly congratulate all those who contributed to this series. I am confident that your work will offer valuable insights for our shared mission—a safer and better-prepared Europe. As the risks we face become ever more complex, we remain committed to deepening partnerships with academic and research institutions, and to promoting life-long learning. Together, we can harness knowledge and innovation to build a resilient Europe—one that is always ready for what lies ahead.

Brussels, Belgium

Hadja Lahbib
European Commissioner for Equality
Preparedness and Crisis Management

Funding Acknowledgements

This publication is part of the Springer book series *Civil Protection, Disaster Management, and Resilience.* The first seven open access volumes were developed within the project: ***EUMA—'Creating a European Higher Education Network for Master's Programmes in Disaster Risk Management' (2024–2025).***

The project has received funding under the **Knowledge for Action in Prevention and Preparedness (KAPP) Call 2023** within the **Union Civil Protection Mechanism (UCPM)** programme of the **European Commission** (Grant Agreement No. 101140384). The publication of this volume has only been made possible through the generous support of this programme.

Within the framework of the EUMA project, a curriculum for a new international master's programme (M.Sc.) has been developed. The postgraduate master programme, entitled '***International Disaster Management and Civil Protection***', will be launched in **October 2026** at the **Postgraduate Center of the University of Vienna**.

Building on the outcomes of EUMA, the follow-up project ***EUMAplus—'Creating Open Access Digital Educational Material in Disaster Risk Management and Civil Protection'*** (2026–2027) is developing open access teaching and learning materials.

Competing Interests The authors have no conflicts of interest to declare that are relevant to the content of this volume.

How to cite this volume:

Sommario E., & Billing, P. (Eds.) (2026): International Legal and Institutional Frameworks for Disaster Risk Management. In: Glade T., Billing P., Bony-Dandrieux A., McMullan C., Savimaa R., Schrott L., Sommario E., van Westen C. J. (Series Eds.): Civil Protection, Disaster Management, and Resilience (Vol. 4), Springer Nature: Heidelberg. https://doi.org/10.1007/978-3-032-11806-6

How to cite a chapter in this volume:

De Siervo, G., & Citossi, F. (2026): Actors and Coordination in Emergencies: A Global Overview. In: Sommario E., & Billing, P. (Eds.): International Legal and Institutional Frameworks for Disaster Risk Management. (Vol. 4), Springer Nature: Heidelberg, pp 1–141.

Acknowledgements

This book is the outcome of a joint research initiative named EUMA (Creating a EUropean Higher Education Network for Master's Programmes in Disaster Risk Management), implemented between 2024 and 2025 by a consortium composed of European universities. We are grateful to the European Commission for its financial support, which made both the EUMA project and the publication of this volume possible.

The authors of the contributions contained in the present volume benefited greatly from the comments and discussions that emerged during the various workshops and intensive courses organised within the EUMA project. We wish to express our sincere gratitude to all participants in these events and to the numerous external colleagues whose insightful suggestions greatly enriched the review process.

We would also like to thank the anonymous reviewers for their careful reading of the manuscript. Their comments and suggestions contributed significantly to the quality and coherence of the volume.

Special thanks are due to Riccardo Luporini and Niamh Reilly for their invaluable assistance in the final stages of preparing this manuscript. Finally, we are grateful to the editorial team at Springer for their professional guidance and support throughout the publication process.

Bonn, Pisa
August 2026

Peter Billing
Emanuele Sommario

Contents

Editors and Contributors

About the Editors

Emanuele Sommario is Associate Professor of International Law at the Scuola Superiore Sant'Anna in Pisa, where he holds the Jean Monnet Chair on *European and International Human Rights Standards in Conflicts and Disasters*. He has published extensively on international disaster law and serves as a Lecturer in the training activities of the European Union Civil Protection Mechanism (UCPM).

Peter Billing Ph.D. is a former Deputy Director for Emergency Management at the European Commission's Directorate-General for European Civil Protection and Humanitarian Aid Operations (DG ECHO). He is currently working as a senior consultant and lecturer for various disaster management institutions and universities in Europe.

Contributors

Peter Billing Department of Geography, University of Bonn, Bonn, Germany

Francesca Citossi Centre for Higher Defence Studies (CASD), Italian Ministry of Defence, Rome, Italy

Giovanni De Siervo Italian Department of Civil Protection, Rome, Italy

Riccardo Luporini Scuola Superiore Sant'Anna, Pisa, Italy

Emanuele Sommario Scuola Superiore Sant'Anna, Pisa, Italy

Abbreviations

AADMER	ASEAN Agreement on Disaster Management and Emergency Response
AHA Centre	ASEAN Coordinating Centre for Humanitarian Assistance
ASEAN	Association of Southeast Asian Nations
BSEC	Black Sea Economic Cooperation
CBRN	Chemical, Biological, Radiological and Nuclear
CDEMA	Caribbean Disaster Emergency Management Agency
CECIS	Common Emergency Communication and Information System
CMCS	Civil Military Coordination Service
CRED	Centre for Research on the Epidemiology of Disasters
DG ECHO	Directorate-General for European Civil Protection and Humanitarian Aid Operations
DRR	Disaster Risk Reduction
ECHR	European Convention on Human Rights
ECPP	European Civil Protection Pool
ECtHR	European Court of Human Rights
EEA	European Environment Agency
EEAS	European External Action Service
EFTA	European Free Trade Association
EM-DAT	Emergency Events Database
EMSA	European Maritime Safety Agency
EMT	Emergency Medical Teams
EMTCC	Emergency Medical Teams Coordination Cell
ERC	Emergency Relief Coordinator
ERCC	Emergency Response Coordination Centre
ESPON	European Spatial Planning Observation Network
EU	European Union
EUCPT	EU Civil Protection Teams
FAO	Food and Agriculture Organization

GHO	Global Humanitarian Overview
HC	Humanitarian Coordinator
HCT	Humanitarian Country Team
HNRP	Humanitarian Needs and Response Plans
IASC	Inter-Agency Standing Committee
ICJ	International Court of Justice
ICRC	International Committee of the Red Cross
IEL	International Environmental Law
IFRC	International Federation of Red Cross and Red Crescent Societies
IHL	International Humanitarian Law
IHR	International Health Regulations
IHRL	International Human Rights Law
ILC	International Law Commission
ILO	International Labour Organization
IMO	International Maritime Organization
INSARAG	International Search and Rescue Advisory Group
IOM	International Organization for Migration
IOs	Intergovernmental Organisations
IPCC	Intergovernmental Panel on Climate Change
KAPP	Knowledge for Action in Prevention and Preparedness
LEMA	Local Emergency Management Authority
MEDEVAC	Medical evacuation
MIC	Monitoring and Information Centre
MSFD	Marine Strategy Framework Directive
NATO	North Atlantic Treaty Organisation
NGO	Non-Governmental Organisation
NIMS	National Incident Management System
OCHA	UN Office for the Co-ordination of Humanitarian Affairs
OECD DAC	Development Assistance Committee of the Organisation for Economic Cooperation and Development
OSOCC	On-Site Operations Coordination Centre
RC	Resident Coordinator
SRSG	Special Representative of the Secretary General
UCPKN	Union Civil Protection Knowledge Network
UCPM	European Union Civil Protection Mechanism
UN	United Nations
UN ISDR	UN International Strategy for Disaster Reduction
UNCT	United Nations Country Team
UNDAC	United Nations Disaster Assessment and Coordination
UNDP	United Nations Development Programme
UNDRO	United Nations Disaster Relief Office

UNDRR	United Nations Office for Disaster Risk Reduction
UNECE	UN Economic Commission for Europe
UNEP	United Nations Environmental Programme
UNFCCC	United Nations Framework Convention on Climate Change
UNGA	UN General Assembly
UNHCR	UN High Commissioner for Refugees
UNITAR	United Nations Institute for Training and Research
WFP	World Food Programme
WHO	World Health Organization

List of Figures

List of Tables

Introduction

1

Peter Billing and Emanuele Sommario

Abstract

This volume surveys contemporary Disaster Risk Management (DRM) through an institutional-legal lens. Drawing on EM-DAT trends, it notes the marked rise in disasters since the twentieth century, the escalating economic losses, and the increased vulnerability driven by climatic, demographic and geopolitical change. The surge in fatalities due to the recent COVID-19 pandemic underscores the need for an all-hazard preparedness approach. Internationally, the UN has shifted from reactive relief to integrated risk governance, anchored in the Sendai Framework (2015–2030) and aligned with the 2030 Agenda and the Paris Agreement. At a regional level, the EU has strengthened the Union Civil Protection Mechanism and is working to improve the speed and effectiveness of its interventions. Normatively, the IFRC's IDRL Guidelines steer domestic facilitation of cross-border aid, and the ILC's Draft Articles on the Protection of Persons in the Event of Disasters furnish a blueprint for a comprehensive treaty on international DRM expected by 2027. This book maps actors and coordination mechanisms and analyses rules on disaster response/recovery and on disaster prevention/preparedness. Conceptually, it adopts UN terminology on hazard, exposure, vulnerability and coping capacity, and explicitly engages the disaster-risk equation linking these variables to explain why hazards become disasters. Finally, it

P. Billing (✉)
Department of Geography, University of Bonn, Bonn, Germany
e-mail: pbillng@uni-bonn.de

E. Sommario
Scuola Superiore Sant'Anna, Pisa, Italy
e-mail: emanuele.sommario@santannapisa.it

© The Author(s) 2026

E. Sommario and P. Billing (eds.), *International Legal and Institutional Frameworks for Disaster Risk Management*, Civil Protection, Disaster Management, and Resilience 4,
https://doi.org/10.1007/978-3-032-11806-6_1

grounds disaster relief in humanitarian principles and links it with the 'do no harm' imperative.

Keywords

Disaster risk management • Sendai framework • Disaster risk equation • International disaster law • Union civil protection mechanism

1.1 Introduction

Virtually every scholarly work on Disaster Risk Management (DRM) begins by highlighting the staggering and increasing number of catastrophic events that have occurred in recent decades, along with their devastating consequences. This volume is no exception, as only by engaging with the empirical data can the magnitude of the problem and the urgency of identifying effective solutions be fully appreciated.

1.1.1 Data and Trends

Even though there are inevitable annual variations, data from the Emergency Events Database (EM-DAT)[1] provide sufficiently robust evidence that globally the number of disasters[2] has indeed significantly increased since the beginning of the twentieth century. Annual figures of natural disaster impacts have oscillated between 40,000 and 50,000 deaths and between 100 and 600 million affected people on average over the last two decades.[3] It is noteworthy that the number of affected people and fatalities caused by natural disasters has decreased over the years, mainly due to progress made in early warning systems, evacuations, improved design of infrastructure and better coordinated disaster response mechanisms. However, figures risk increasing due to new hazards and vulnerabilities caused by climatic, demographic, economic and geopolitical changes. In terms of economic damage, for instance, the decadal average damage caused by natural disasters, measured in percentage of global gross domestic product (GDP), has more than doubled since the 1960s.[4] In absolute terms, economic damage caused by natural disasters in the

[1] The database is run by the Centre for Research on the Epidemiology of Disasters (CRED) at the Catholic University of Louvain, www.emdat.be. See also https://ourworldindata.org/natural-disasters. These, like all other online resources referenced in the volume, were last accessed in June 2026.

[2] The database defines a 'disaster' as an event that meets one of the following criteria: 10 or more fatalities; 100 or more affected people; a declaration of state of emergency; or a call for international assistance. For more insights on the definitions and their limitations, see https://doc.emdat.be/docs/.

[3] Actual figures vary according to source and period selected. UNDRR estimated 42,000 fatalities per year from 2015 to 2022, see https://www.undrr.org/explainer/uncounted-costs-of-disasters-2023.

[4] Ritchie et al. (2022).

period 1980–2024 was around 7 trillion United States Dollars (USD),[5] an amount more or less equivalent to the combined annual GDP of the United Kingdom (UK) and France.[6] If one considers not only natural disasters, but also recent technological and public health disasters, the decade 2015 to 2024 witnessed a dramatic increase in both the number of fatalities and those injured compared to the previous decade: an increase of 249% and 173%, respectively.[7] This is mainly attributable to the COVID-19 pandemic and the effects of climate change. In Europe, heat waves have become the most lethal disaster.

1.1.2 Actors and Mechanisms

Heightened awareness of the plight of disaster victims has drawn attention to the importance of adequate national and international rules and structures for DRM. International actors have not remained idle and have sought to adapt their roles and functions to better address these challenges. Since the early 2000s, the United Nations (UN) has progressively expanded and refined its disaster response framework, shifting from largely reactive, humanitarian-centred approaches to a broader, integrated disaster risk management system closely linked to development, resilience-building, and climate adaptation. Notable milestones include the adoption of the Sendai Framework for Disaster Risk Reduction (2015–2030), which shifted the focus from managing disasters to managing disaster risk and emphasised the importance of prevention, preparedness, response, and recovery. At the same time, the United Nations Office for Disaster Risk Reduction (UNDRR) emerged as a critical actor in supporting implementation, monitoring, and reporting under Sendai. More recently, disaster response has been increasingly integrated with the 2030 Agenda for Sustainable Development, the Paris Agreement on Climate Change, and the Agenda for Humanity. In addition, the COVID-19 pandemic underscored the need for all-hazard preparedness, blending biological disaster response with traditional disaster management approaches.

Regional DRM systems also witnessed an unprecedented development. Europe strengthened its regional disaster management by enhancing cooperation, coordination, and rapid response capacities. The European Union Civil Protection Mechanism (UCPM) was expanded and reformed to offer better disaster response both within and outside EU borders. Major developments came in 2013, with the adoption of Decision 1313/2013/EU, emphasising risk assessment, disaster preparedness, and mutual assistance, as well as

[5] Munich Re (2024).

[6] Statista, The 20 countries with the largest gross domestic product https://www.statista.com/statistics/268173/countries-with-the-largest-gross-domestic-product-gdp/.

[7] Own calculations by the authors using the EM-DAT database, with data extracted on 14 April 2025. The World Health Organization (WHO) estimates the COVID-19 has killed more than seven million people, see WHO COVID-19 dashboard, https://data.who.int/dashboards/covid19/deaths.

the creation of the Emergency Response Coordination Centre (ERCC), offering enhanced 24/7 operational support.

1.1.3 Legal Developments at the International Level

The international legal landscape has also changed at an unprecedented rate. A first significant effort to organise international rules on disaster response was led by the International Federation of Red Cross and Red Crescent Societies (IFRC) through its International Disaster Response Laws, Rules and Principles (IDRL) Programme, launched in 2001. The IFRC developed the non-binding Guidelines for the Domestic Facilitation and Regulation of International Disaster Relief and Initial Recovery Assistance (IDRL Guidelines), offering technical advice for States to legally prepare for international disaster assistance. In parallel, a more traditional codification effort was undertaken by the UN International Law Commission (ILC). In 2007, the ILC began its work on the topic 'Protection of persons in the event of disasters' and finally produced a set of 18 Draft Articles (DAs), covering the disaster response, recovery, and prevention phases. The text has been identified as the blueprint for a future, comprehensive treaty on disaster risk management, scheduled for adoption by States by the end of 2027. This Treaty will represent a significant contribution to the development and consolidation of international disaster law, harmonising what is currently a disparate landscape.

This volume seeks to provide readers with a comprehensive overview of the issues outlined above. While it does not purport to offer an exhaustive treatment of the subject, it is intended to serve as a reference text for those interested in gaining a deeper understanding of the institutional and legal mechanisms underpinning contemporary DRM efforts at the international level. To this end, the book is structured into two parts, each comprising two chapters. The first part examines the principal institutions and actors involved in DRM activities. In particular, the Chapter 2 will present the traditional and new players in the international humanitarian arena, as well as the most important coordination mechanisms that have been developed in recent years. The Chapter titled 3 will then zoom in on the most advanced example of a regional mechanism aimed at disaster management, namely, the UCPM. It will also provide an insight into the features of the civil protection structures in EU Member States and in other states participating in the UCPM. The second part analyses the legal framework governing DRM activities, roughly dividing them between rules applying to the post-disaster and pre-disaster phases. The Chapter titled 4 will address the norms pertaining to disaster response and recovery activities. After reviewing some key definitions, it will focus on the typical obligations codified in disaster response instruments and then offer a description of the emerging legal framework concerning recovery activities. The Chapter on 5 will do the same with respect to disaster prevention and disaster preparedness measures and highlight the role of disaster risk reduction (DRR) in the broader disaster risk management framework.

1.1.4 The Contemporary Notion of *Disaster* and some Relevant Definitions

Before proceeding, it is necessary to clarify a few preliminary concepts that are relevant to all four chapters of the volume. These include the conceptualisation of the notion of *disaster*, the categorisation of activities aimed at dealing with disasters in various phases that compose the so-called *disaster risk management cycle*, and the humanitarian principles that should inform disaster response efforts.

Starting with how calamities are commonly addressed in academic studies, it is becoming common knowledge that a disaster is not merely the materialisation of a hazardous event, but needs to be conceptualised considering its concrete effects on the affected communities. Indeed, natural or technological hazards may become disasters depending on the exposure and vulnerability of the affected community, as well as its capacity to react. According to this approach, disasters are not isolated incidents but rather events that expose the underlying dysfunctions of the affected society. The position that 'there is no such thing as a natural disaster'[8] has therefore become a *Leitmotiv* in disaster studies. This approach has also been expressed through an equation seeking to capture the distinguishing elements of disaster risk[9]:

$$\text{Disaster risk} = \frac{\text{Hazard (H)} \times \text{Exposure (E)} \times \text{Vulnerability (V)}}{\text{Capacity to cope (C)}}$$

While agreeing on shared definitions of the terms employed in the equation is a rather difficult endeavour, for the purposes of the present volume we will adopt those identified by the Open-Ended Intergovernmental Expert Working Group on Indicators and Terminology Relating to Disaster Risk Reduction,[10] which can be considered to reflect the approach adopted by the UN. In its 2016 report, the Working Group defines the term *hazard* as '[a] process, phenomenon or human activity that may cause loss of life, injury or other health impacts, property damage, social and economic disruption or environmental degradation'.[11] Hazards may originate from natural, anthropogenic, or socio-natural causes. Natural hazards are primarily linked to natural processes and phenomena, such as earthquakes, floods, and volcanic eruptions. Anthropogenic hazards arise entirely or predominantly from human activities and decisions, and include phenomena

---

[8] Hartman and Squires (2006).

[9] For a thorough explanation of how the equation has emerged and how it is conceptualised, see IPCC (2012, p. 69 and ff).

[10] The Working group was established by the UN General Assembly (UNGA) through its resolution 69/284, 3 June 2015, and was mandated with the development of a set of possible indicators to measure global progress in the implementation of the Sendai Framework for Disaster Risk Reduction 2015–2030, with the support of the UNDRR.

[11] UNGA, Report of the Open-Ended Intergovernmental Expert Working Group on Indicators and Terminology Relating to Disaster Risk Reduction, UN Doc. A/71/644, 1 December 2016, p. 18.

such as industrial pollution, nuclear radiation, toxic wastes, dam failures, factory explosions, fires, and chemical spills.[12] Finally, socio-natural hazards have their origin in a combination of natural and human-induced factors, such as environmental degradation and climate change.[13]

Exposure is defined as '[t]he situation of people, infrastructure, housing, production capacities and other tangible human assets located in hazard-prone areas'.[14] If a hazardous event occurs in an area where there is no exposure (meaning no people, property, or assets are present) then no risk is generated. However, exposure patterns are dynamic and can evolve over time as a result of factors such as population growth, economic development, and urbanisation, all of which can increase the concentration of people and assets in hazard-prone areas, thereby elevating disaster risk.

Vulnerability is defined as '[t]he conditions determined by physical, social, economic and environmental factors or processes which increase the susceptibility of an individual, a community, assets or systems to the impacts of hazards'.[15] It is a multifaceted concept, shaped by a range of physical, social, economic, and environmental factors, which encompasses the broader structural and contextual conditions that constrain the capacity of people and communities to withstand, respond to, and recover from hazardous events.[16]

Lastly, closely connected to vulnerability is the notion of *coping capacity*, which is defined as 'the ability of people, organisations and systems, using available skills and resources, to manage adverse conditions, risk or disasters'.[17] It requires sustained awareness, resources, and effective management, both under normal circumstances and during emergencies. Strengthening coping capacities plays a critical role in reducing disaster risk, as it enhances the capacity to absorb, adapt to, and recover from the impacts of hazardous events.

---

[12] Note that this category does not encompass the occurrence or risk of armed conflict or other forms of social instability or tension, which are governed by international humanitarian law, international human rights law and national legislation, ibid.

[13] Although we fully endorse the shift in terminology from 'natural disasters' to 'natural-hazard-induced disasters', the former term will occasionally appear in this volume. This is partly because many documents and laws have long classified disasters as 'natural' or 'technological'; those labels cascade into reporting, insurance, emergency declarations, and research trends, so changing them has been slow. In addition, 'natural disaster' is a compact, familiar phrase: it compresses a complex causal chain into two words, aiding public understanding.

[14] UNGA, Report of the Open-Ended Intergovernmental Expert Working Group, p. 18.

[15] Ibid., p. 24.

[16] Adams et al. (2022, pp. 13–29).

[17] UNGA, Report of the Open-Ended Intergovernmental Expert Working Group, p. 12.

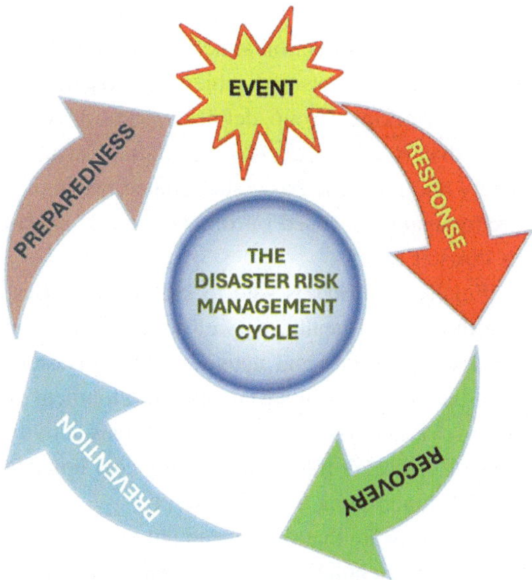

Fig. 1.1 The disaster risk management cycle. *Source* Developed by the authors

1.1.5 The DRM Phases

In making sense of the many activities linked to DRM, one of the most useful organising principles is the disaster cycle paradigm. This model delineates a set of strategies for managing disaster risk and is commonly considered to include four phases: prevention, preparedness, response, and recovery.[18] While more specific definitions are provided in the various chapters, it is worth quickly presenting the meaning attributed to these different terms by the UN Working Group (Fig. 1.1).

The *prevention* phase includes '[a]ctivities and measures to avoid existing and new disaster risks'.[19] Disaster prevention aims at completely avoiding the potential adverse impacts of hazardous events. Where certain disaster risks cannot be eliminated, prevention aims at reducing vulnerability and exposure so that the risk of disaster is minimised. *Preparedness* refers to '[t]he knowledge and capacities developed by governments, response and recovery organisations, communities and individuals to effectively

[18] Note that *prevention* is often associated with, and sometimes replaced by, *mitigation*. The two terms are closely connected, as both refer to measures aimed either at eliminating the risk of disasters (prevention) or at minimising their harmful effects when they cannot be entirely avoided (mitigation). For the sake of simplicity, we will use the term *prevention* to refer to all activities designed to proactively reduce the risk of disasters. Sometimes, these two concepts are grouped together under the broader umbrella of DRR activities. See UN ISDR (2004), p. 17.

[19] Ibid., p. 21. Examples include dams or embankments that eliminate flood risks, land-use use regulations that do not permit any settlement in high-risk zones, seismic engineering designs that ensure the survival and function of a critical building in any likely earthquake and immunization against vaccine-preventable diseases.

anticipate, respond to and recover from the impacts of likely, imminent or current disasters'.[20] It includes action aimed at building the capacities needed to efficiently manage all types of emergencies and achieve orderly transitions from response to sustained recovery.

The Working Group defines *response* as '[a]ctions taken directly before, during or immediately after a disaster in order to save lives, reduce health impacts, ensure public safety and meet the basic subsistence needs of the people affected'. Disaster response is predominantly focused on immediate and short-term needs and is sometimes referred to as disaster relief.[21]

Finally, *recovery* is said to include activities aimed at 'the restoring or improving of livelihoods and health, as well as economic, physical, social, cultural and environmental assets, systems and activities, of a disaster-affected community or society, aligning with the principles of sustainable development and "build back better", to avoid or reduce future disaster risk'.[22] The term includes activities aimed at the reconstruction or restoration of critical infrastructures, services, housing, facilities, and livelihoods required for the full functioning of a community.

We are aware that the conceptual model of the disaster cycle has come under increasing criticism.[23] Scholars and practitioners have pointed out that representing disaster-related activities as circular can be misleading, as it implies a static, repetitive sequence rather than progressive or evolving processes. Moreover, the boundaries between the phases are becoming increasingly blurred, especially in the context of slow-onset events, such as those linked to climate change, which unfold gradually over time and do not fit neatly into distinct temporal phases.[24]

However, using the disaster cycle as a framework has the advantage of putting the response phase, typically at the forefront, into context and highlights the interconnections between risk management strategies. For instance, failures to prevent disaster risk often amplify the impact of the later disaster, turning an otherwise manageable event into a more serious risk to life or property. Adopting the disaster cycle paradigm is also useful because it simplifies the phases of DRM in a way that is easily relatable in terms of a sequence of events that need to happen after or before a calamity strikes. Yet it should be emphasised that prevention, preparedness, response, and recovery are to be considered as overlapping activities, not separate stages.

---

[20] Ibid. Examples of preparedness activities and include contingency planning, the stockpiling of equipment and supplies, the development of arrangements for coordination, evacuation and public information, and associated training and field exercises.

[21] Ibid., p. 22. Typical response measures include the activation of counter-disaster systems; search and rescue activities; the provision of emergency food, shelter, medical assistance; evacuation measures; surveying and assessing buildings and other infostructure in the area affected by the disaster.

[22] Ibid., p. 21.

[23] Richardson (2005, pp. 48–50).

[24] Staupe-Delgado (2019, p. 629).

1.1.6 The Core Humanitarian Principles

Since the core of disaster management activities still lies in the response phase,[25] reference must be made to the basic principles that should inform any type of humanitarian action, including disaster relief. These are the humanitarian principles of *humanity*, *impartiality,* and *neutrality*. Many add as a fourth principle *independence*, which requires humanitarian actors to be autonomous, i.e. not subject to control, subordination, or influence by political, economic, military, or other non-humanitarian objectives. However, some consider that independence is of limited import if applied to national civil protection actors, which are traditionally closely linked to the respective governmental authorities.[26] The conformity of the aid offered with humanitarian principles is provided by a plethora of instruments dealing with humanitarian assistance, applicable in armed conflicts and/or in disaster contexts, such as GA Resolution 46/182 and the IFRC's IDRL Guidelines. Although Resolution 46/182 did not elaborate on these principles, the UN Office for the Co-ordination of Humanitarian Affairs (UNOCHA) developed meaningful definitions.[27]

The principle of *humanity* demands that human suffering must be addressed wherever it is found, with particular attention to the most vulnerable populations such as children, women, and the elderly. The dignity and rights of all victims must be respected and protected. *Impartiality* requires that humanitarian actions must be carried out on the basis of need alone, giving priority to the most urgent cases of distress and making no distinctions based on nationality, race, gender, religious belief, class, or political opinions. According to the principle of *neutrality*, humanitarian actions must be provided without engaging in hostilities or taking sides in controversies of a political, religious, or ideological nature.

While the principles of humanity and impartiality clearly apply across all disaster contexts, neutrality might, at first glance, appear to be a context-specific principle, inherently linked to situations of conflict. Nevertheless, in the context of disaster response, neutrality should be interpreted as requiring that 'those responding to disasters should abstain from any act which might be interpreted as interference with the interests of the State'.[28] In other words, neutrality obliges assisting actors to refrain from taking sides in controversies of a political, racial, religious, or ideological nature.

---

[25] Indeed, we will occasionally use the expression *disaster management* in lieu of *disaster risk management* when the discussion focuses on *response* to natural or technological hazards. However, we recognize that disaster risk management is the umbrella term covering the entire, risk-focused cycle. In this volume, *disaster management* refers to the operational subset that addresses the event and its consequences.

[26] Perchinig et al. (2017, p. 18). However, when UCPM assets contribute to humanitarian assistance efforts promoted by the EU, the principle of independence applies to them too, see Decision 1313/2013/EU, Article 26.3.

[27] Guiding Principles for Public-Private Collaboration for Humanitarian Action, prepared by the World Economic Forum and UNOCHA, December 2007.

[28] International Law Commission, Third report on the protection of persons in the event of disasters, UN Doc A/CN.4/629, 31 March 2010, para. 29.

Of course, humanitarian principles are not uncontested.[29] However, adherence to their spirit is fundamental for securing and sustaining access to affected populations, whether in the context of natural disasters, armed conflicts, or complex emergencies. Although often regarded as abstract or aspirational ideals, humanitarian principles constitute a practical and indispensable framework for fostering trust, legitimacy, and acceptance among all stakeholders. Even more importantly, they are recognised as key operational tools by virtually all humanitarian actors.[30]

Lastly, reference should be made to another core tenet of contemporary humanitarian action, the so-called 'do no harm' principle.[31] Adherence to this principle requires that when humanitarian actors intervene to help people affected by crises (including disasters) they must be careful not to worsen existing problems or create new harms through their actions.

Key Sentences

- **The Intensifying Disaster Risk Landscape**—Despite improvements in preparedness and response, disasters are becoming more complex and costly due to climate change, urbanisation, and emerging vulnerabilities, resulting in growing economic and societal impacts.
- **Disasters as Products of Structural Vulnerability**—Disasters arise from the interaction between hazards, exposure, vulnerability, and coping capacity, revealing underlying social and institutional weaknesses rather than constituting purely natural events.
- **Shift from Reactive Response to Risk Management**—Contemporary disaster governance prioritises prevention, preparedness, recovery, and resilience, embedding disaster risk management within broader sustainable development and climate adaptation frameworks.

29 See, e.g. Slim (1997, p. 342). The principles have also been challenged by the introduction of new frameworks advocating for an increased integration and coordination between humanitarian assistance, development assistance, and peace (the s.c. 'triple nexus'); see Brown et al. (2024).

30 The European Union, for instance, put them at the core EU of their flagship document on humanitarian aid, see European Consensus on Humanitarian Aid, Joint Statement by the Council and the Representatives of the Governments of the Member States meeting within the Council, the European Parliament and the European Commission, 2008/C 25/01, 30 January 2008, paras 10–14.

31 The principle was developed by influential American scholar and practitioner Mary B. Anderson in the early 1990s, see Anderson (1999). Since then, it has transformed in a tool that underpins many operational frameworks. See, for instance, Sphere Association. The Sphere Handbook: Humanitarian Charter and Minimum Standards in Humanitarian Response, fourth edition, Geneva, Switzerland, 2018.

- **The Progressive Consolidation of International and Regional Frameworks**—International and regional actors have progressively institutionalised disaster governance through coordinated mechanisms and emerging legal frameworks, enhancing collective preparedness and response capacity.
- **The Central Role of Humanitarian and 'Do No Harm' Principles**—Humanitarian principles remain essential operational tools, ensuring access, legitimacy, and effectiveness of assistance, while the 'do no harm' principle mitigates the risk of exacerbating existing vulnerabilities.

References

Adams R et al (2022) Social vulnerability and disasters: development and evaluation of a CONVERGE training module for researchers and practitioners. Disaster Prev Manag 31:13–29

Anderson MB (1999) Do no harm: how aid can support peace–or war. Lynne Rienner Publishers

Brown S et al (2024) The peace dilemma in the triple nexus: challenges and opportunities for the humanitarian–development–peace approach. Dev Pract 34:568–584

Hartman C, Squires GD (eds) (2006) There is no such thing as a natural disaster: race, class, and hurricane Katrina. Routledge

Intergovernmental Panel on Climate Change, IPCC (2012) Managing the risks of extreme events and disasters to advance climate change adaptation. A special report of working groups I and II of the intergovernmental panel on climate change. Cambridge University Press

Munich Re (2024) Natural disasters worldwide. https://www.munichre.com/en/risks/natural-disasters.html

Perchinig B et al (2017) Humanitarian aid and civil protection policies in the European Union and the MICIC agenda. International Centre for Migration Policy Development. https://micicinitiative.iom.int/resources/humanitarian-aid-and-civil-protection-policies-european-union-and-micic-agenda

Richardson BK (2005) The phases of disaster as a relationship between structure and meaning: a narrative analysis of the 1947 Texas City explosion. Int J Mass Emerg Disasters 23:27–54

Ritchie H, Rosado P, Roser M (2022) Natural disasters. https://ourworldindata.org/natural-disasters

Slim H (1997) Relief agencies and moral standing in war: principles of humanity, neutrality, impartiality and solidarity. Dev Pract 7:342–352

Staupe-Delgado R (2019) Progress, traditions and future directions in research on disasters involving slow-onset hazards. Disaster Prev Manag 28:623–635

UN International Strategy for Disaster Reduction, UN ISDR (2004) Living with risk: a global review of disaster reduction initiatives. United Nations

Open Access This chapter is licensed under the terms of the Creative Commons Attribution 4.0 International License (http://creativecommons.org/licenses/by/4.0/), which permits use, sharing, adaptation, distribution and reproduction in any medium or format, as long as you give appropriate credit to the original author(s) and the source, provide a link to the Creative Commons license and indicate if changes were made.

The images or other third party material in this chapter are included in the chapter's Creative Commons license, unless indicated otherwise in a credit line to the material. If material is not included in the chapter's Creative Commons license and your intended use is not permitted by statutory regulation or exceeds the permitted use, you will need to obtain permission directly from the copyright holder.

Actors and Coordination in Emergencies: A Global Overview

2

Giovanni De Siervo and Francesca Citossi

Abstract

This chapter explores the complex and evolving landscape of actors and coordination mechanisms in international emergency response. As crises become more frequent, diverse, and interconnected, the number and nature of humanitarian actors—ranging from national governments and international organisations to NGOs, civil society, private companies, and military forces—have multiplied, challenging effective coordination. The chapter examines the legal and operational frameworks governing national responses and the mechanisms for requesting and managing international assistance. It analyses global coordination systems led by the UN, such as the cluster approach and UNDAC, while special attention is paid to civil-military cooperation. The authors highlight the need for stronger preparedness, standardisation, and accountability at all levels, with an emphasis on the participation of affected communities and the effective application of the 'do no harm' and 'last resort' principles.

Keywords

Disasters • Relief • International aid • Coordination • Humanitarian actors • United Nations • NGOs • Military

---

G. De Siervo (✉)
Italian Department of Civil Protection, Rome, Italy
e-mail: Giovanni.DeSiervo@protezionecivile.it

F. Citossi
Centre for Higher Defence Studies (CASD), Italian Ministry of Defence, Rome, Italy
e-mail: francesca.citossi@unicasd.it

© The Author(s) 2026
E. Sommario and P. Billing (eds.), *International Legal and Institutional Frameworks for Disaster Risk Management*, Civil Protection, Disaster Management, and Resilience 4,
https://doi.org/10.1007/978-3-032-11806-6_2

2.1 Defining the Context[1]

The past decades have witnessed a worldwide expansion in the devastating consequences of natural and human-made disasters, alongside the emergence of other types of crises at the global level.[2] On one hand, the frequency and intensity of natural disasters are increasing and, on the other hand, the reappearance of emergencies which were considered relegated to history—as well as the substantial increase in new forms of crises linked to hybrid threats—are forcing countries and international organisations to reassess their capability to manage such situations and to update emergency procedures. Recent experiences, from COVID-19 to the Türkiye and Syria 2023 earthquake, have shown that the disruptive effects of serious emergencies undermine even well-organised systems and require mutual aid among countries with the support of international institutions and civil society organisations.

In the first instance, in order to assess whether the available instruments are adequate and sufficient, it is necessary to determine the responsible authority in charge of managing the consequences of an extreme event. According to the UN Charter, it is clear that no Member State should theoretically have the right to intervene in the territory of another country without consent. Therefore, each country has the right—or indeed the duty—to organise its disaster and crisis management system according to its own culture, history, and social structure.

As a consequence, different organisational methods have developed over the years in different countries, depending on the type of emergencies a given territory is most prone to. There are systems that focus on natural disasters, while others focus more on human-made crises or possible military attacks. Depending on social organisation and different traditions, countries have chosen more centralised or decentralised systems.[3]

The evolution and interconnection of the different sectors within modern societies have led to the need to involve more governmental and civil society entities in the management of the consequences of an event, so today the trend is increasingly towards whole-of-government and whole-of-society response systems.[4] As such, the supply chain or the communication systems are indeed almost always managed by private companies.

In principle, the country should be independent in identifying risks and threats affecting its territory, and it should structure its own system to manage them accordingly. However, in cases in which the affected state should deem that its national response capacities are overwhelmed, it can call for international assistance, defining the way and timing as well as the operational modalities by which aid should be managed. This is not always the case due to several factors. First, when a country requests assistance, the severity of the event does not allow it to be handled using national resources only. This type of situation is

[1] Part of this chapter was originally published by Giovanni De Siervo: De Siervo (2012).

[2] See Humanitarian Action, Global Humanitarian Overview (2025).

[3] For more details, see Billing's chapter in the present volume.

[4] See Niinistö (2024).

symptomatic of a difficult situation in which those in charge of managing the emergency are likely to be working under enormous pressure, both personal and social, which might affect their capability to act calmly and according to established rules. Furthermore, the political optics in these situations plays a very substantial role in deciding whether to request assistance and how to manage the emergency. In this respect, recent developments in social media communications have introduced another complicating factor, influencing public opinion substantially and much faster than traditional media. This challenge is leading to a rethinking of reference standards, tools, and ways of working by states to communicate both risks and crisis situations.

Moreover, in recent decades, the number and kind of actors involved in emergency response at the international level has increased substantially, raising growing concerns on the coordination and effectiveness of the system. Another element that complicates the response system is the type of crisis to deal with. Traditionally, a distinction was made between natural disasters and human-induced or complex emergencies.[5] Today, the scope of crises is even more diverse, especially due to extreme weather events affecting areas that were previously considered safe. The reappearance of geopolitical tensions, coupled with technological improvements, has substantially increased the risk of hybrid and cyberattacks as additional challenges.

In this chapter, we will mainly deal with the management of the consequences of emergencies regardless of their triggering factor. We will adopt an *all-hazard consequence management approach,* without investigating crisis drivers and their management systems, some of which aim to resolve the root causes of emergencies and conflicts. We will explore the main characteristics of the most relevant actors and coordination systems with a view to offering a nuanced portrait of the complexity and the challenges of the contemporary humanitarian landscape.

2.1.1 National Frameworks and Challenges[6]

Article 2 of the UN Charter defines and recognises the principle of sovereignty. Every country, therefore, is, in principle, in charge of emergency management on its own territory. In this regard, the 'landmark' is the 1991 UN General Assembly (UNGA) Resolution 46/182 which defines the main principle underlying coordination efforts: '*Each State has the responsibility first and foremost to take care of the victims of natural disasters and other emergencies occurring on its territory. Hence, the affected State has the primary role in*

[5] Complex emergencies, as defined by the Inter-Agency Standing Committee (IASC), are severe humanitarian crises with significant breakdown of authority due to conflict, requiring international response beyond single-agency capacity, involving massive civilian suffering, displacement, security risks for aid workers, and complex coordination challenges, see IASC (1994).

[6] For additional elements about the role of the disaster-affected State in managing the emergency, see Sommario's chapter in the present volume.

the initiation, organisation, coordination, and implementation of humanitarian assistance within its territory'. Indeed, the first responders are always the locals, and national emergency facilities and the affected population are always the first to provide relief support in an emergency. The responsibility for organising its own national response system is therefore clear, and each country develops it according to historical, political, and cultural traditions, as well as its social structure.

There are systems in which military structures are heavily involved and others in which civil society organisations are the key actors. Countries with a decentralised administrative structure normally choose to differentiate between territorial levels based on the principle of subsidiarity, while other, more centralised countries opt for more unified, top-down systems. Countries with territories at high risk of natural hazards usually have major systems for monitoring risks and threats and, where possible, engage in extensive contingency planning activities by providing stockpiles and procedures in case disaster strikes. On the contrary, nations whose primary concern is a possible conflict will concentrate their resources on countermeasures to a possible attack and on the preservation of societal functions under any circumstances.

The criteria applied to face risks, threats, foresight, and assessment, as well as their results, are critical because they have a direct impact on how consequence management systems are built. Prioritising either natural hazards or security threats will affect the countries' prevention and preparedness choices and therefore their capability to respond to an emergency or to mitigate its effects. These systems are named differently according to their general objective: there are civil defence, civil protection, disaster management, or contingency agencies.[7] Although there are some sector-specific examples, there are no internationally unambiguous definitions of the roles and responsibilities of these organisations. However, regardless of their names and specific national focuses, these are all structures that oversee the management of the consequences of disasters and crises on the population.

Each organisational system has its own features: some divide the responsibilities among the different organisations involved with strong coordination, while in others, single entities are in charge of the whole crisis management, covering both operational and strategic aspects. Those organisations are also positioned differently in their respective governments; in some cases, they depend on the ministries for internal affairs, the ministries of security, the prime minister, or the president. Their main task is to organise and implement relief operations on their national territory. Their action is mainly focused on coordinating the action of all the different actors involved: ministries, administrations, and civil society. Each country has selected its coordination methodology, hierarchical or consensus-based, on the basis of sectors/clusters or competences. However, in general, it is possible to identify at least the highest coordination authority mandated by the national government to ensure the response operations.

[7] Alexander (2002).

In 1971, the UNGA '*invite[d] potential recipient Governments [t]o appoint a single national disaster relief coordinator to facilitate the relief of international aid in times of emergency*'. When the severity of the disaster overwhelms the national response capacities and there is a need for an international response, states have, in principle, the responsibility of requesting and coordinating the assistance in their own country. In this case, it is crucial to identify in a very short time who has the mandate for requesting assistance. Foreign relief organisations, both governmental and NGOs, and Civil Society Organisations (CSOs)[8] are willing to intervene in a timely manner for a number of reasons. Timing is crucial for life-saving activities, public attention is highly attuned in the first post-disaster phases, and the '*humanitarian arena*' is very crowded. Furthermore, resources, especially of a technical nature, need time to become operational and often emergency funds can be released only after the event. Foreign agencies need to receive the request as soon as possible, and the affected state needs to conduct a proper needs assessment, which is key for an effective response. Ideally, the disaster would strike a country with a stable government that can assess the damages, call for specific international assistance, and define the framework and rules to follow to provide support, however the reality is often quite different. Even in stable countries, the weight of public opinion can affect even strong governments.

In principle, there is therefore no doubt that coordination responsibility falls on the national governments. The international tools and methodologies would only support national efforts. However, very often states lack sound legal and institutional arrangements for the coordination of international actors and the negative consequences of this preparation gap have been evident in recent operations. Furthermore, the national institutions in charge of coordinating post-disaster operations usually must contend with a very challenging environment: infrastructures are often seriously damaged; coordinators are often personally affected by the event; and intense scrutiny from public opinion and the media is constant. Moreover, national authorities and international relief organisations often have limited knowledge and understanding of each other's complexity, culture, policies, procedures, and working mechanisms. In this complex scenario, the massive influx of international actors and means may pose an additional burden on local capabilities. For these reasons, the international community, over the years, has developed a non-binding set of tools, codes, guidelines, standard operating procedures, and rules to facilitate the role of the affected state.

Traditionally, the national responders belonged to emergency management authorities, public institutions, and civil society, while at international level the Red Cross and Red Crescent Movement, some NGOs, the UN System, and a few other organisations were operational. Today, the number and kind of actors have drastically increased; diversified; and have different missions, methodologies, and agendas.

---

8 For the differences and similarities between NGOs and CSOs please refer to Varga (2020).

2.1.2 The 'Other' Actors

Traditionally, international disaster response has been ensured by agencies depending either on states or international organisations (IOs). At the same time, traditional humanitarian organisations such as the Red Cross and Red Crescent Movement have been providing post-disaster assistance, as have some NGOs. All operated according to widely recognised principles of humanity, neutrality, and impartiality and the more debated principle of independence.[9] The main difference among the actors was linked to their status as governmental or non-governmental organisations. However, in recent years, the number and variety of actors in the humanitarian arena has significantly increased[10] making the coordination task even more complex. Non-state actors (CSOs, private companies, media, etc.) are increasingly present in post-disaster situations and the number of NGOs has drastically increased at a global level.

The typology of actors has a substantial impact on operational modalities, intervention time, and coordination. State actors whose task is to manage the safety and security of their target populations—both in the host and offering countries—have predictable tools, budgets, and procedures, and move according to a logic of inter-state relations. Entities that are regularly engaged in humanitarian work (IOs or INGOs) have capacities, which are also structured and professional, and can operate for longer periods of time in operations abroad because they are structured for this purpose. To operate in a given crisis, however, they need to obtain specific funding either through participation in international appeals or by approaching their own donors. Local civil society organisations (NGOs or CSOs) usually have limited operational and financial capacity but accurate knowledge of the local context, which makes them indispensable for effective and sustainable activities.

Against this backdrop, recent decades have witnessed a return to the centrality of the state in disaster and emergency relief. This situation reflects the changing geopolitical context and an emerging perception that international humanitarian assistance is no longer the preferred option for affected states. Certainly, in complex emergencies, as well as in post-disaster situations, the affected states tend to reaffirm their national sovereignty and to play a more direct role in relief operations. In some instances, they reduce the humanitarian space by limiting the access of international relief organisations and imposing strict controls over their activities. At the international level, governments have also increased their direct presence by using their emergency services or their military resources. Several authors have recognised a growing trend to submit the provision of humanitarian support to political considerations.[11]

An element that is becoming increasingly central in the management of emergencies is the constant involvement of the private sector, both as national and international actors. While the latter have over the years developed capacities to contribute to emergency

[9] Please refer to the introduction of this volume.

[10] Müller-Stewens et al. (2019, pp. 1–7).

[11] Global Humanitarian Platform (2010), Kent (2011).

management based on the principle of corporate social responsibility and the spirit of solidarity, some cases have also emerged over the years of companies interested in intervening after disasters for business purposes. In parallel, the ever-increasing interconnection of the economic and social sectors of modern societies and the handing over of economic sectors by states to private actors have indeed made the presence of private actors in the management of emergencies indispensable. For instance, modern supply chain and telecommunication activities are almost completely managed by the private sector. The proliferation of actors, in permanent competition for funding and visibility, has led several authors to draw parallels between the relief sector and marketplace dynamics.[12] The affected government, state agencies, IOs, international and local NGOs, CSOs, private companies, and military forces have their own features, *modus operandi*, funding mechanisms, time span, and mission; however, all of them are present in the aftermath of the disaster and they are all there to support the same affected population. Therefore, it is crucial to understand how each of these entities can and should work together. The need to give greater importance to the interests of the final beneficiaries and their communities when deciding on priorities and how to deliver assistance has emerged very strongly in recent years. With this in mind, the humanitarian sector has launched a major revision that includes a substantial increase in cash-based assistance systems and structured mechanisms to involve local communities in identifying priorities and actions.[13] UNGA Resolution 46/182 aimed to improve coherence and coordination of aid, encouraging both affected and assisting states to channel offers and assistance requests through the UN Emergency Relief Coordinator, however this has not always been the case. While the beneficiaries in any post-disaster operation must be the local affected population, communities, and institutions, there are growing concerns that relief organisations prioritise accountability to their donors above local actors.

The member governments of the Development Assistance Committee (DAC) of the Organisation for Economic Cooperation and Development (OECD)[14] continue to contribute to the humanitarian and post-disaster response, and they remain the major donors at the international level.[15] However, non-DAC government contributions (mostly BRICS[16] and Gulf States) have recently increased considerably[17] as well as donations from corporate, philanthropic, or public actors. The presence of new donors who are not part of the

[12] See Polman (2010), Lilly and Bowden (2024).

[13] Refer to *The Grand Bargain* to explore the ways used to include the interests of local communities: https://interagencystandingcommittee.org/grand-bargain.

[14] For 50 years now, the OECD DAC has grouped the world's main donors, defining and monitoring global standards in key areas of development. More info on the OECD is available at http://www.oecd.org/

[15] For a detailed analysis of the changing landscape of donors in humanitarian action, see Macrae (ed) (2002) and Development Initiatives, DI (2024).

[16] BRICS is an intergovernmental organisation comprising ten countries: Brazil, Russia, India, China, South Africa, Egypt, Ethiopia, Indonesia, Iran, and the United Arab Emirates.

[17] For an analysis of current trends: Pearson (2024).

traditional discussion or coordination constituencies may create additional challenges. Private philanthropy, non-DAC donors, and public fundraising are today increasingly actively present on the stage and coordination of their action presents a challenge.

Another implication is that different actors are not always adequately prepared, or comprehending of the risks and implications of their actions, and are not always aware of the agreed standards and procedures. Nowadays, there is an almost unanimous acceptance of the humanitarian principles, however in practice these are implemented differently by different actors. Every organisation has its own mandate, mission, and operational methodology that deeply affects the manner in which the principles are implemented, and the operations which are carried out.[18] In this respect, it is interesting to notice the number of different coordination constituencies for NGOs. Furthermore, some NGOs or entities do not commit themselves to participating in coordination arrangements.[19] There are no possible legal instruments to stop any group of people from intending to create an entity and to access the humanitarian marketplace. The only effective check that can be implemented by the affected government is by setting up domestic legal barriers for the provision of disaster assistance (i.e. access, visa, customs, registration, labour law, professional liability, etc.).[20]

The use of state entities raises very serious, and long-debated, concerns related to the nature of the different actors, especially the military. By their own nature, these assets depend on a government, calling into question the humanitarian principle of independence. Furthermore, the military follows its own reporting line and a separate chain of command that might exclude them from coordination. The UN 'integrated mission'[21] concept has faced serious questions regarding the risk of submitting humanitarian action to political action, undermining the importance of independence/impartiality to political considerations. The presence of so many actors raises the fundamental question of how to ensure that assistance can reach those in need and support the authorities that requested

[18] The Feinstein International Center of the Tufts University has recognised the following Typology of Humanitarian Actors: '*"Principled-Dunantist"* based on the basic tenets of humanitarianism developed by Henri Dunant. Proponents of principle-centred action argue for a narrower definition of humanitarianism limited to life-saving assistance and protection of civilians, based on core principles of neutrality, impartiality, and independence. *"Pragmatist-Wilsonian"* recognise the importance of principles but place a higher premium on action, even when this means putting core principles in jeopardy. They identify broadly with foreign policy objectives of their home government, whose funds they often use. *"Solidarist"* beyond the provision of assistance and protection they address the root causes of conflict, which are political at the core. *"Faith-based"* the world's major religious traditions, western and non-western alike, embody humanitarian affirmations and obligations. In addition to international faith-based agencies that do not usually engage in proselytising, there is a wide variety of religious organisations at the local level. Faith-based entities may themselves embody principled, pragmatist, and/or solidarist features'. See Donini et al. (2008, p. 11).

[19] For a comprehensive analysis of this topic refer to ICVA (2024).

[20] See Sommario's chapter in the present volume.

[21] Hofman (2014).

help in an effective manner, while respecting humanitarian principles and the principle of '*do no harm*'.

2.2 The Coordination Mechanisms

The issue of coordination in emergency situations widely differs according to the kind of crisis that has to be dealt with. Traditionally, two kinds of emergencies generally require a large-scale international relief operation: natural or human-made disasters and '*complex emergencies*'. In the first case, the affected government has the primary responsibility to manage the relief activities, it has to 'call' the crisis and to ensure coordination; the international actors are expected to act in accordance with its indications while respecting its sovereignty. In case of a '*complex emergency*'[22] the humanitarian agencies may assume direct protection responsibilities of the affected population, due to the breakdown of national institutions, and the coordination mechanisms are mainly internationally driven. Some authors argue that responding to a natural disaster is very different from operating in complex emergencies and slow-onset crises, not only due to the emergency environment but also because of differing logistical considerations.[23]

Coordinating disaster response in the aftermath of a natural or human-made disaster is, usually, less problematic than in a complex emergency. The general principles, criteria, and tools are the same. However, the dynamics of emergency coordination are intrinsically connected to the type and origin of the disaster, its location and the political, economic, administrative, and cultural situation of the affected country. Successes and failures of coordination depend greatly on the specific situation of each affected country. The presence of a stable national authority, with a proper administrative structure and an organised response system, would certainly simplify the provision of international support as well as its coordination. On the other hand, political tensions, social unrest, or conflicts would raise major concerns, notably because international humanitarian response to disasters must be provided according to the recognised humanitarian principles.

In particular, the case of a sudden-onset disaster, such as earthquakes or floods, coordination raises different concerns compared to a slow-onset crisis (such as droughts), at least initially. Complex emergencies or slow-onset crises are indeed the result of structural problems that need to be addressed through different management methodologies. Large sudden-onset disasters overwhelm the normal situation in a matter of hours. For this reason, the response takes place in a unique, high-pressure emergency environment

[22] A complex emergency is 'a humanitarian crisis in a country, region or society where there is total or considerable breakdown of authority resulting from internal or external conflict and which requires an international response that goes beyond the mandate or capacity of any single agency and/or the ongoing United Nations country program', Inter-Agency Standing Committee, IASC (2004, p. 5).

[23] Katoch (2006, p. 155).

and aims to save as many lives as possible. In these instances, relief activities generally involve straightforward operations, such as search and rescue; triage; medical treatment; and provision of water, food, and shelter. Local authorities must respond in a very short timeframe, and they are often overwhelmed by emerging needs that push them to require international assistance. Clearly, the coordination settings in these situations need to be as specific as possible.

2.2.1 The Local Emergency Management Authority

National governments are usually reluctant to admit they do not have the capacity to cope with the effects of a catastrophe. Therefore, open requests for assistance do not happen often,[24] or are made late for political reasons. More frequently, the requests are transformed into welcoming external offers of assistance, as was the case with India after the 2001 Gujarat earthquake or the 2008 Nargis cyclone in Myanmar.[25]

For a country requesting assistance, there are a number of options: activation of the bilateral channel, a multilateral institution (global or regional), or a combination of the different schemes. The request timing has a direct impact on the coordination efforts since countries in sudden-onset disasters are often not able to prepare a proper needs assessment but end up compiling a standard '*shopping list*' of ordinary emergency items. There is then a supply-driven response, where international actors do not offer supplies and assets according to an agreed needs assessment, but according to the availability in stocks causing gaps and overlaps. The affected country usually does not have the capability to refuse offers because it receives pressure to accept them both domestically and internationally.[26] Secondly, in the aftermath of a major disaster the workload required to scrutinise all possible offers is often well beyond domestic capacity. IOs and foreign governments normally deal with affected governments through official offers and requests prior to providing assistance, but this is not the common practice of other international actors, such as NGOs.

The ability to coordinate requires good skills, a sound knowledge of the applicable legal framework, of the international relief actors, and of their dynamics. Each state should prepare for the possibility of receiving incoming assistance, defining dedicated legal and operational frameworks. However, this fails to happen in the majority of cases. Regional

---

[24] ALNAP (2010, p. 22).

[25] After the Gujarat Earthquake the Indian government publicly welcomed 'acts of solidarity'; in the aftermath of Nargis Cyclone, the government of Myanmar welcomed international humanitarian assistance, see The Tripartite Core Group (2008).

[26] Regardless of the fact that Italy did not request international assistance in the aftermath of the 2009 L'Aquila earthquake, there were several pressing offers of international assistance coming from various countries and NGOs. Nonetheless it was officially reported that international offers would be considered only after a proper needs assessment, some NGOs decided anyway to deploy their teams to Italy.

organisations such as the European Union have invited their members to implement dedicated legislation. The '*Council Conclusions on Host Nation Support*'[27] call on the 'Member States and the Commission to step up structured work in the field of Host Nation Support' and to 'make the necessary arrangements in the existing organisational structure in order to either receive assistance or facilitate the transit of assistance'.[28] Similarly, the '*Guidelines for the domestic facilitation and regulation of international disaster relief and initial recovery assistance*'[29] (the 'IDRL Guidelines'), adopted by the state parties to the Geneva Conventions at the 30th International Conference of the Red Cross and Red Crescent Movement on 30 November 2007, are meant to assist governments to be better prepared for the common legal problems connected to international response operations. However, so far very few countries have developed their own legal and organisational framework.

Once the offer is accepted, the incoming assistance should be channelled through the existing coordination authorities. The system set up by the national government to manage the relief operations is commonly called the Local Emergency Management Authority (LEMA). As per the overall national systems, each country develops it according to its own national organisational culture and methodologies.[30] National coordination settings should ensure that national and international responses are complementary. In the US, for example, the National Incident Management System (NIMS) is the emergency management doctrine applicable at all administrative levels and across functional disciplines to ensure effective resource coordination. It is based on enhanced cooperation and interoperability among responders through a standardised, flexible, all-hazards approach to incident management, recognising a clear chain of command. The NIMS Incident Command System became a reference model at international level. It consists of a standard management hierarchy and procedures for managing temporary incidents of any size with a clear command structure. The LEMA should set up dedicated procedures and tools to coordinate international actors and to make the best possible use of incoming resources. This would also facilitate national ownership in terms of a coordinated response. International actors would be required to refer to the national coordination settings. The practice of recent operations, however, has demonstrated that some international actors deliberately bypass national coordination structures[31] and fail to inform domestic authorities of their activities. Concurrently, it is still rare to find countries that have established a dedicated coordination system for international aid. Furthermore, there is often a change of

---

[27] 305th Justice and Home Affairs Council meeting, Brussels, 2 and 3 December 2010.

[28] The Conclusions on HNS were followed by the Commission Recommendation on EU Host Nation Support Guidelines (29 November 2024).

[29] Endorsed by the UNGA in 2008 Res. 63/139, Res. 63/141, and Res. 63/137, see IFRC (2007, 2008).

[30] In Italy the Augustus methodology and DiComAC concept were developed. OECD (2010).

[31] ALNAP (2010, p. 23).

coordination settings in the various phases of the response, from the acute emergency period to the recovery/development phases, which may present challenges.[32]

Very often, capacity shortcomings—particularly at the national government level—significantly constrain effective coordination. For example, during the 2010 Haiti earthquake, not only were those in authority not able to produce a proper needs assessment, but they did not have the capacity to maintain their coordination role.[33] In this case, the international community needed to support the affected country by lending immediate capacity and technical assistance.

2.2.2 The United Nations Global Coordination

Based on the assumption that the host country is the authority in charge of managing the emergency, any coordination by the international community must be done in close cooperation with the local authorities. Within the United Nations, which also has the task of facilitating mutual support among states for humanitarian purposes, the Emergency Relief Coordinator (ERC) oversees all crises requiring humanitarian assistance by facilitating coordination among all agencies involved. The ERC has the task of advising the Secretary General on strategic choices that require coordinated action.

The ERC also holds the role of Under-Secretary General for Humanitarian Affairs and is the authority responsible for acting as a point of contact for all actors involved in the humanitarian response to an emergency. In post-disaster assistance, the ERC is operating mainly through their country-level representatives, the 'humanitarian coordinators'. It relies on the Office for the Coordination of Humanitarian Affairs (OCHA), created within the UN Secretariat as a part of the 1998 reform.[34] OCHA is not directly operational (and is therefore seen as more impartial), but it must '*[m]obilize and coordinate effective and principled humanitarian action in partnership with national and international actors in order to alleviate human suffering in disasters and emergencies; Advocate the rights of people in need; Promote preparedness and prevention; Facilitate sustainable solutions*'.[35] Another important instrument coordinated by the ERC is the IASC, created in 1991, which brings together senior representatives of all UN agencies and major humanitarian bodies. This forum has a role in coordination, policy development, and decision-making.[36]

To ensure coherence of action and coordination among the different actors in a specific country, the Resident Coordinator (RC) and the United Nations Country Team (UNCT)

---

[32] Bennett et al. (2006, p. 40).

[33] Grünewald, Binder (2010, p. 42).

[34] UNGA, Res A/51/1950, Renewing the United Nations: A programme for Reform, Report from the Secretary General.

[35] OCHA Mission Statement. http://www.unocha.org.

[36] IASC. https://interagencystandingcommittee.org/the-inter-agency-standing-committee.

system have been established. The United Nations RC is the highest United Nations official and the chief of the UN diplomatic mission in a country, except when there is an appointed Special Representative of the Secretary General (SRSG). The RC has the same rank as an ambassador, is the designated Representative of the UN Secretary General, and leads the UNCT.[37] The function is usually performed by the Resident Representative of the United Nations Development Programme (UNDP).[38] In case of an emergency, if international humanitarian assistance is required and a separate Humanitarian Coordinator (HC) position is not established, RCs are often called to assume the HC[39] role and coordinate country-level response to humanitarian crises. The HC plays a crucial role in ensuring effective coordinated humanitarian action at country level, particularly in the case of natural disasters. The system has a solid normative framework, developed and implemented at inter-agency level and strengthened by recent reforms.

The UNCT includes all the entities of the UN system that carry out operational activities for development, emergency, recovery, and transition.[40] It is composed of all the UN entities that carry out operational activities in a specific country, and it is the forum that ensures inter-agency coordination and decision-making at national level. The UNCT is led by the RC. Its decision-making processes are participatory, collegial, based on consensus and mutual accountability, and its decisions are non-binding. The main purpose of the UNCT is to facilitate all involved agencies to plan and work together to ensure the delivery of tangible results in support of the development agenda. They share assessments, define a common working programme, and jointly mobilise resources to ensure a coherent action of the different UN entities. All UNCT members have direct-line accountability to their own organisation, as well as collegial accountability to the RC.

The coordination system of international actors is based on the basic humanitarian principles and rules defined by the IASC and ERC. With the activation of the HC function,

---

[37] The RC system aims at bringing together the different UN agencies to improve the efficiency and effectiveness of operational activities at the country level. It includes all organisations of the UN system dealing with operational activities for development, regardless of their formal presence in the country. At country level, RCs are tasked to ensure leadership within the UN system, facilitating the coordination and coherence of all UN operational activities, advocating and promoting the UN global agenda. The UNGA established the concept of a single official to coordinate operational activities within the UN system in 1977 (GA Resolution 32/197). The concept has evolved, and the role of the UN Country Team has been reaffirmed in several occasions.

[38] The fact that the RC assumes more than one role at a time is criticised. For this reason, it is often stressed that the RC should be independent from any agency, including his/her parent entity, a neutral position *vis à vis* the government, strong humanitarian experience, and a mix of diplomatic and operational skills.

[39] The designation is decided by the UN ERC in consultation with the IASC. In this function, the RC/HC is in charge for supporting national efforts and he/she is accountable to the ERC for the strategic and operational coordination of the response of the UNCT member agencies and relevant humanitarian actors.

[40] The UNCT exists in 136 countries, covering all of the 180 countries where there are UN programmes.

the Humanitarian Country Team (HCT) is also established. This brings together all the UN agencies, and all the major humanitarian agencies and organisations in the country. At the national level, the HCT is the highest level of coordination of humanitarian bodies in the country and the National Cluster also participates in this team when activated. It is a strategic level that must give guidance to all the different levels of coordination that are activated. It is a strategic decision-making forum, so the number of actors represented is limited, and it has a structure that follows that of the IASC plus any other donors or entities particularly active in the country.[41] The main task of the HCT is to coordinate the humanitarian action in the country and to make sure that it is provided according to humanitarian principles. It operates in support and coordination of local authorities. It does not replace the UNCT, and the two entities coexist because their focus is different.

2.2.2.1 The Clusters

A system composed of multiple autonomous bodies—each dependent on voluntary contributions—lacks strategic leadership and is liable to create gaps and a lower preparedness level. Building on the lessons identified from the review of emergencies in which the international response had failed to demonstrate efficiency and effectiveness, in 2005 the ERC launched the Humanitarian Reform to address the shortcomings that had emerged. The review highlighted gaps and inconsistencies in the humanitarian response system, especially related to predictability and coordination, and suggested methods to address them. The three main innovations are linked to the cluster approach, a more predictable financing and strategic leadership.

While financial innovations may play a central role in coordination because they provide adequate, flexible, and predictable resources,[42] as well as strengthening the HC function, the more crucial innovation for coordination was the institution of the cluster approach. Starting from the unpredictability of humanitarian responses, the foundational idea is to address the issue by identifying sectoral coordination at a global level with a clearly identified lead organisation. Following the recommendations, the IASC originally identified nine global clusters with their respective lead organisation. Today, the number of clusters has increased and includes the following areas: Food security/WFP-FAO, Nutrition/UNICEF, Water sanitation and hygiene/UNICEF, Health/WHO, Shelter/IFRC-UNHCR, Camp coordination and management/UNHCR-IOM, Shelter UNHCR-IFRC, Protection/UNHCR, Logistics/WFP, Emergency telecommunications/WFP, Education/UNICEF-Save the Children, and Early recovery/UNDP (Fig. 2.1).[43]

---

[41] For more information, https://interagencystandingcommittee.org/sites/default/files/migrated/2019-02/hct_tors.pdf.

[42] For an in-depth analysis of the HCP and the different available tools, https://knowledge.base.unocha.org/wiki/spaces/hpc/pages/3993075713/About+the+HPC.

[43] For more details, https://reliefweb.int/topics/cluster-coordination.

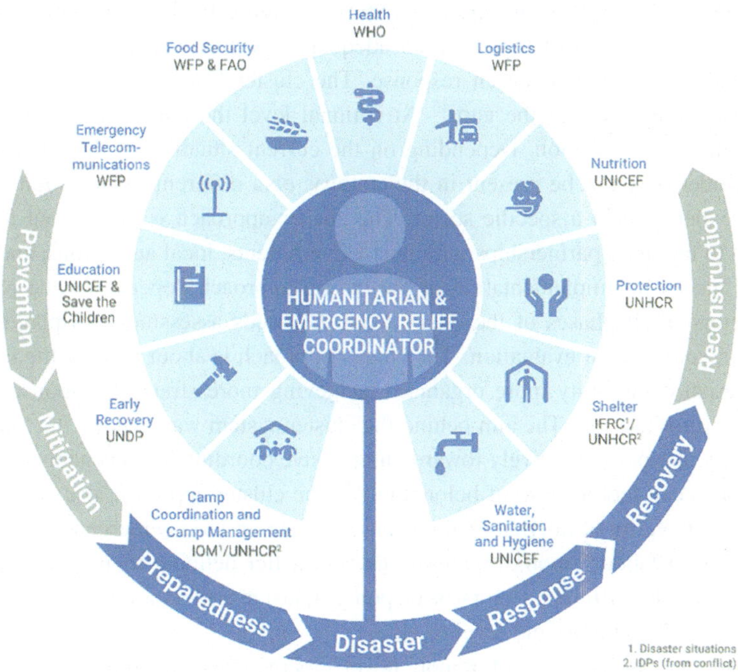

Fig. 2.1 The humanitarian cluster system. *Source* www.unocha.org/we-coordinate

The aim of the cluster approach is to ensure predictable and timely response in crisis. It seeks to improve the partnerships among the different actors (UN agencies, the International Red Cross and Red Crescent Movement, international organisations, and NGOs[44]) to make sure that sufficient capacity at global level is available in all the main sectors/ areas of response.

The principle of designating a lead organisation at global level is meant to address the long-lasting issue of sectoral leadership. If it does not imply the existence of a chain of command, the global cluster lead organisation is mandated to ensure that response capacity is ready and that assessment, planning, and response activities are carried out in collaboration with partners in accordance with agreed standards and guidelines. It also makes the international humanitarian community a better partner for host governments by identifying a focal point for the specific sector. Furthermore, the cluster leader is also expected to act as the 'provider of last resort'.

[44] The International Committee of the Red Cross (ICRC), like some NGOs (following the Dunantist approach), do not take part in the cluster approach but they coordinate with UN to achieve efficient response for the affected people. However, at the global level they participate as observers in many of the cluster working group meetings.

The ‘provider of last resort’ concept represents, theoretically, a commitment of sector leaders to do their utmost to ensure an adequate and appropriate response where there are critical gaps in humanitarian response. The cluster leader in those cases may need to commit itself to filling the gap.[45] At national level the cluster leader does not need to be a global organisation. Depending on the current situation of the country, a global cluster leader may not be present in the country, or a different organisation may have a specific experience in a specific sector. The cluster approach is also meant to provide a platform to facilitate partnerships with host Governments, local authorities, and local civil society. Successful implementation of the cluster approach depends on the participation of all actors in all phases of the crisis, from the needs assessment, to planning, implementation, and ex-post evaluation. The cluster approach is about making the international humanitarian community more organised, achieving more strategic responses, planning, and better prioritisation. The aim behind the cluster system was making the ‘humanitarian community to work collectively toward an inclusive coordination mechanism to which all stakeholders can feel a sense of belonging’.[46] The cluster approach also serves to ensure better accountability in international responses to humanitarian emergencies, by clarifying the division of labour among organisations, and better defining their roles and responsibilities within the different sectors of response. Cluster leaders are accountable, at global level, to the ERC and at field level to HCs.

In principle, in the case of major, novel emergencies that require a multi-sectoral response with participation of a wide range of international humanitarian actors, the cluster approach should be applied from the beginning to plan and organise the international response. The activation must be requested to the ERC by the HC after consultation with relevant partners. The ERC in coordination with Global cluster leaders ensures agreement at global level and communicates the agreement to HC/RC and partners within 24 hours. The HC/RC will then inform the host government and all partners. The cluster approach should be implemented in close cooperation with the host government. The cluster leaders at country level ensure the development and the maintaining of proper links with all relevant local stakeholders. At the global level the cluster system is accompanied by a consolidated humanitarian appeal process to mobilise resources collectively in what is now called the Global Humanitarian Overview (GHO). The GHO is made up of country-level Humanitarian Needs and Response Plans (HNRP) which in turn consolidate the individual humanitarian projects of agencies according to sectors.

[45] Regrettably, none of the UN agencies were willing to accept the financial and reputational risk of taking on the role of provider of last resort, and the concept has virtually been abandoned. See Mark Bowden and Angela Penrose, ‘The evolution of the humanitarian marketplace and NGO financing models’, unpublished Working Paper available at https://www.nuffield.ox.ac.uk/media/5567/the-evolution-of-the-humanitarian-marketplace-and-ngo-financing-models.pdf, p. 9.

[46] Adinolfi et al. (2005, p. 47).

Since its establishment, the cluster approach has been applied in several emergencies with mixed results. It has shown good potential to improve not only international coordination but also coordination between the international community and local stakeholders. All evaluations of the cluster approach have shown that it did work in protracted crises and slow-onset emergencies to fill gaps and to avoid duplication of effort, as well as management of the newcomers. In sudden-onset disasters, the effectiveness of clusters appears to be limited. The system needs time to be established and a complete analysis of local structures and capacities should be undertaken beforehand. The fact that the cluster system was 'designed' as an interim solution but is now being used as a system in itself raises problems because clusters are process—rather than action-oriented. Their effectiveness depends on the motivation and ability of the appointed lead agencies to put the resources and effort into making their cluster work, as well as on the willingness of being coordinated by all different agencies. Furthermore, in their initial implementation, clusters largely excluded national and local actors. They are coordination settings established for foreign agencies where the use of the English language does not facilitate the integration with the affected state's coordination and response mechanisms. Through the innovations of *The Grand Bargain*[47] major efforts were put in place to bring the cluster closer to local communities and actors.

The leadership aspect most crucially differentiates the cluster approach from earlier models of coordination. Having a designated lead agency with responsibility not only for the performance of its own programme but for the entire sectoral response represents the most important benefit of this approach. Whenever more than one cluster is activated, inter-cluster coordination, managed by UN OCHA, takes care of the operational level to limit gaps and duplications. These forms of coordination must not neglect the national settings and actors, and the activation of the cluster system must be performed by HCs in consultation with local authorities. The different local ministries should also co-chair the different sectoral clusters. With the international community becoming aware of the need to bring coordination and decision-making closer to the local beneficiaries, efforts have been undertaken in recent years to increasingly involve local communities. Initiatives have been taken to decentralise coordination centres to listen to the needs of communities (area-based coordination).

In recent years, the humanitarian sector has embarked upon a profound renewal. Following the 2016 World Humanitarian Summit, a shared pathway was launched by the world's key players in the sector, aiming to find answers to the growing needs registered at a global scale. Among the various measures agreed during the Summit was *The Grand Bargain,* which included 51 commitments to improve the effectiveness and accountability of the humanitarian system, with a renewed focus on communities directly affected

[47] The Grand Bargain, launched during the World Humanitarian Summit in Istanbul in May 2016, is a unique agreement between some of the largest donors and humanitarian organisations who have committed to get more means into the hands of people in need and to improve the effectiveness and efficiency of the humanitarian action, https://interagencystandingcommittee.org/grand-bargain.

by crises and emergencies. This new approach was changed with *The Grand Bargain 2.0* (2021) which also includes a focus on more support and funding tools for local and national responders. In 2023, *The Grand Bargain 3.0* was launched: it aims at increasing the involvement of affected communities in decision-making on interventions that directly impacts them. Within this framework, UN OCHA also launched a readjustment of the principles governing the coordination of humanitarian action by adopting the OCHA's Strategic Plan 2023–2026.[48]

2.2.2.2 The United Nations Disaster Assessment and Coordination (UNDAC)

Lifesaving activities, typical within the first phases of sudden-onset disasters, must be implemented in a very short time, especially as media exposure may induce the influx of a multitude of actors on site of the disaster. For this reason, other coordination tools were developed in the framework of OCHA.

Established in 1993, the UNDAC[49] is a system created to fill in the possible coordination gaps in the first days after the disaster. It consists of four different elements: Staff, Methodology, Mobilisation Procedures and Operational Partnerships and Equipment. UNDAC is a roster of very experienced emergency managers available to intervene in real time. The roster is managed by OCHA and is composed of professional staff coming from over 40 countries either from national emergency services or international humanitarian organisations. UNDAC teams are regionally organised to make sure that their deployment takes place in a short period of time and may take advantage of their knowledge of the region and specific risks affecting the area. They are fully equipped and deployable in 12–24 hours in every part of the world. The UNDAC team is available free of charge to the affected country to support the coordination during the initial phases of the disaster through the provision of technical services under the leadership of the RC/HC. The UNDAC team supports the affected country by providing a coordination platform able to support a multi-sectoral assessment and a proper information management. It is activated on request of the affected government or the UN Country team. UNDAC teams are in charge for establishing and managing the On-Site Operations Coordination Centre (OSOCC). This concept was developed by the International Search And Rescue Advisory Group (INSARAG) guidelines to facilitate coordination of Search And Rescue teams[50] that need a special support. Usually, the SAR teams work side by side with UNDAC teams since they are deployed in the earliest phases of a disaster. Since its creation, the OSOCC concept proved to be an effective tool to facilitate the coordination of international resources in the early phases of a sudden-onset disaster (Fig. 2.2).

---

[48] OCHA's Strategic Plan 2023–2026, https://www.unocha.org/publications/report/world/ochas-strategic-plan-2023-2026-transforming-humanitarian-coordination.

[49] For more information, see UNDAC (2024).

[50] UNGA Res 57/150 (2002).

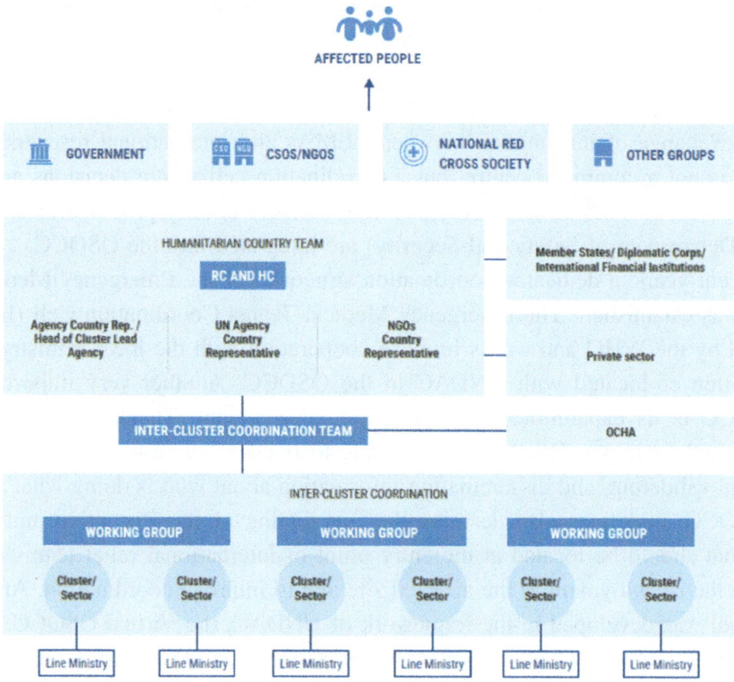

Fig. 2.2 Humanitarian coordination structure. Humanitarian coordination at the country level. *Source* UNDAC Handbook (2024)

The OSOCC guidelines,[51] though mainly dedicated to UNDAC, provide guidance to the first agency arriving on the site of the disaster and requested to set up a coordination structure in coordination with OCHA. It is a rapid response tool that cooperates with LEMA in the immediate aftermath of a sudden-onset disaster. It is inspired by the Incident Command Centers[52] commonly established at national level. As a rapid response tool, to be effective OSOCC must be established in the immediate aftermath of the disaster, to be operational simultaneous to the arrival of the first international responders and it is expected to be active until national authorities or traditional UN structures can take over the coordination activities. OSOCC's objectives are: to act as a link between international responders and government of the affected country; to provide a system for coordinating

[51] OSOCC (2018).

[52] 'Incident command system provides a flexible core mechanism for coordinated and collaborative incident management, whether for incidents where additional resources are required or are provided from different organisations within a single jurisdiction or outside the jurisdiction, or for complex incidents with national implications', US National Incident Management System, https://www.fema.gov/emergency-managers/nims.

and facilitating the international relief activities; and to provide an information management platform. OSOCC responds to a series of concerns: it is a single point of entry for the LEMA managers, it offers a focal point for the international community, it provides the physical space and facilities to ensure the essentials for coordination and facilitates, and the exchange of information between LEMAs and international rescuers. However, OSOCC is not a command centre, but a coordination cell where decisions are taken by consensus. To be more effective, the common services usually provided by the UN (e.g. the UN Department of Safety and Security) are integrated into the OSOCC.

In recent years, a dedicated coordination structure for the Emergency Medical Teams (EMT) was established. The Emergency Medical Teams Coordination Cell (EMTCC) is managed by the WHO and works in close cooperation with the local Ministry of Health and is often co-located with UNDAC in the OSOCC. Another very important feature of OSOCC is its capabilities in information management. Thanks to its coordination platform and tools, the OSOCC team is able to build a common situational picture by gathering, validating, and disseminating information about who is doing what and where. The OSOCC guidelines also describe the functioning of the Reception and Departure Centre that should be located at the entry point of international relief teams in order to facilitate their deployment to the affected site and to improve coordination. Another very useful tool was developed in the framework of UNDAC: the Virtual OSOCC. This web-based platform provides real-time information management dedicated to relief workers. It is accessible to all relief workers and every user can upload information and resources. It provides an overall picture of the activities of all the different actors that decide to utilise it.

2.2.2.3 UN Integrated Missions

Integrated missions have emerged as a development of the UN involvement in Peace Support Operations.[53] They are called upon to manage 'complex emergencies' that simultaneously call on the political, military, humanitarian, and developmental remits of the UN system. The concept can be defined as an approach through which the UN seeks to support countries to move from a situation of complex emergency to a lasting peace, by using the system-wide available response, within an overall political-strategic management framework. It applies to missions headed by a SRSG with the overall responsibility of the whole operation, covering both the military and humanitarian/development dimensions. In these instances, the Humanitarian Coordinator's role is given to the SRSG or to a Deputy SRSG. Such missions institutionalise a permanent and highly visible link between the political instances (SRSG and RC) and the HC. The integrated missions model raised serious criticism and objections from the humanitarian community, which

[53] Peace Support Operations: Peace Building, Peace Keeping, Peace Making, Peace Enforcing.

argued that it weakens the perception of the UN as neutral and impartial. In those situations, humanitarian action is characterised as just another tool to reach the long-term political objective.[54]

2.2.3 Regional Mechanisms

In parallel with the global mechanisms, several dedicated coordination tools were developed at regional level, for example, in Europe, Latin America, Asia, and among NATO countries.

The European Union (EU) over the years has developed several different tools to intervene in emergency situations.[55] Moreover, EU member countries are among the more active countries in international post-disaster operations. For these reasons, the issue of coordination assumes a special meaning, in particular after the innovations of the Lisbon Treaty[56] which, in order to answer the increasing demand for the EU to become more coherent (especially at international level), established the European High Representative of the Union for Foreign Affairs and Security Policy and the European External Action Service (EEAS). For emergency preparedness and response purposes, as well as to ensure coordination, the Union Civil Protection Mechanism (UCPM) has been established in 2001 and today is managed by the Directorate-General for European Civil Protection and Humanitarian Aid Operations (DG ECHO) of the European Commission. It connects the civil protection authorities of 27 EU Member states and 10 additional participating states. It has at its disposal the Emergency Response Coordination Centre (ERCC) and has become the cornerstone of the collective response capacity of the EU. The UCPM is fully integrated into the broad international coordination ecosystem, and in case of emergencies in Europe, it is expected to be the first international coordination mechanism to support the affected state.

In Latin America, different sub-regional mechanisms were established: the *Comité Andino para a Prevenção e Atençãoaos Desastres* (CAPRADE); the *Centro de Coordinación para la Prevención de los Desastres Naturales en América Central* (CEPREDENAC); the *Agencia Caribeña para el Manejo de los Desastres y la Emergencia* (CDEMA); the *Regional Group on Risks, Emergencies and Disasters for Latin America and the Caribbean* (REDLAC); and the *Reunión Especializada de Reducción de Riesgos de Desastres Socionaturales, la Defensa Civil, la Protección Civil y la Asistencia Humanitaria*

---

[54] Some examples of UN integrated missions are the United Nations Integrated Mission in Timor-Leste (UNMIT), which supported the country's transition to independence, and the United Nations Multidimensional Integrated Stabilization Mission in Mali (MINUSMA), focused on security, political processes, and state authority in Mali.

[55] For more details, see Billing's chapter in the present volume.

[56] Treaty of Lisbon amending the Treaty on the European Union and the Treaty establishing the European Community (1 December 2009).

del MERCOSUR (REHU). Each of them is provided with specific tools and procedures. To ensure action coherence and to improve the synergetic approach and exchange of information among the various bodies active on these issues in Latin America, a special forum has been created: the *Foro de coordinaciòn y cooperacion de los mecanismos Sub-regionales de Gestiòn del Riesgo de Desasters de la Amerìca.*

In Southeast Asia, the *Association of Southeast Asian Nations* (ASEAN) *Agreement on Disaster Management and Emergency Response* (AADMER) has a structured coordination system based on the ASEAN Coordinating Centre for Humanitarian Assistance (AHA Centre)[57] and is able to deploy teams of experts (the Emergency Response and Assessment Team—ERAT) to support affected countries. In recent years also the *Southern African Development Community* (SADC) developed a capacity to strengthen coordination for preparedness and response in the region.

The *North Atlantic Treaty Organisation* (NATO) plays a role both as regional organisation and—due to the specific nature of the organisation—in facilitating civil-military coordination.[58] NATO established the *Euro-Atlantic Disaster Response Coordination Centre* (EADRCC) located at NATO Headquarters, which is the focal point for information sharing among NATO countries. In close consultation with UN-OCHA, it coordinates the responses of member countries and ensures close liaison with the organisations involved in international disaster response.

NATO's involvement in post-disaster assistance follows a few basic principles. The affected country remains the responsible party for disaster management, while the UN retains the primary role in the coordination of international disaster relief operations. The EADRCC's role is coordination rather than direction, while NATO members are committed to coordinate with the UN and to observe the Guidelines on the Use of Military and Civil Defence Assets in Disaster Relief ('Oslo Guidelines'). The EADRCC does not dispatch coordination teams to the affected country. In the case of a disaster requiring international assistance, it is the prerogative of each nation to decide whether to provide assistance or not.

2.2.4 Working with the Military

Humanitarian organisations, state security forces, and non-state armed groups increasingly operate in the same space during disaster relief operations.[59] The coexistence of large complex organisations in the same operational environment increases the level of complexity, with sometimes poorly coordinated efforts on a common set of objectives.

---

[57] The AHA Centre is an intergovernmental organisation, established by the ten ASEAN Member States with the aim of facilitating cooperation and coordination on disaster management among ASEAN Member States, see https://ahacentre.org/about-us/

[58] See infra para 2.4.

[59] See UN OCHA, Civil-military coordination, https://www.unocha.org/civil-military-coordination

Civilian and military actors operate under different mandates, strategic interests, principles, roles, doctrines, value systems, and operational modalities, sharing a temporal space that requires communication, interaction, and coordination. Over the years, efforts have been made to define an international framework that clarifies the modes of interaction.[60] An effective dialogue among these entities[61] is essential to gain and maintain humanitarian access to the people in need, while at the same time clarifying the lines of command and the division of specialised work. Different scenarios must be considered on a case-by-case basis, remembering that in every situation it is important to maintain a clear separation between the roles of the military and the humanitarian actors, distinguishing their respective spheres of competence and responsibility.[62]

Effective collaboration is the result of two elements: functionality and trust. It might happen that lack of familiarity leads to distrust and the refusal to work with certain partners, which was the case for some time in the Civil-Military (Civ-Mil) relationship,[63] due to different structures, missions, organisational cultures, personnel skills and preparation, procedures, and resources. The nature of interaction between humanitarian and military actors will be different in each operational context and depend on the type of emergency and the specific mandate of deployed military forces. A cornerstone of Civ-Mil guidance remains the Principle of Last Resort,[64] according to which 'assets should be requested only where there is no comparable civilian alternative and only the use of military or civil defence assets can meet a critical humanitarian need', in accordance with internationally agreed Civ-Mil guidelines and recommended practices. It is essential to retain the overall civilian-led, principled nature of the humanitarian response that should be guided by the core principles such as impartiality, neutrality, and independence. However, perfect operationalisation of these principles is rarely achieved in practice. The *do no harm* principle guides decision-making for humanitarians. Some situations require capabilities available only from the military community such as strategic air and sealift, medical support and medical evacuation capacities, and specialised engineering capabilities. Exceptionally, assistance by the military might also be required in order to create appropriate safety conditions for humanitarian workers to deliver aid and operate in complex emergencies.

A fundamental concern of humanitarian organisations is that sharing any information with military actors might violate humanitarian principles and place them at greater risk of attack. The fundamental difference between military and humanitarian aid must be kept in mind: if both provide relief to people in need, military units most of the time cannot

---

60 The original version of this paragraph is part of a paper originally presented by Daniela Irrera and Francesca Citossi at SISP, *Società Italiana di Scienza Politica* in Trieste, 12–14 September 2024 with the title '*Civil-military relations in humanitarian emergencies: the Italian experience*'. Many thanks to Professor Irrera for allowing us to re-elaborate this part for the objectives of this piece of work.

61 Bell et al. (2019, pp. 263–278).

62 Grace et al. (2023).

63 Boin and Bynander (2014).

64 Alejandria et al. (2022, pp. 12, 112, 113).

do it in a neutral and impartial manner. Armed forces serve strategic interest, such as winning the hearts and minds of local communities[65]: therefore, activities carried out in support of the military mission are by definition not 'humanitarian'.

Humanitarian organisations draft their own policies on how to interact with armed actors—state or non-state, military, police, or private security contractors—recalling the main internationally recognised guidelines.[66]

The EU promotes coordination between civilian and military actors in emergencies[67] through Humanitarian Civil-Military Coordination (CMCoord). The scope and form of the Civ-Mil interaction always depend on the specific context of the crisis and take multiple forms: coexistence, consultation, de-confliction, coordination, and cooperation. A case-by-case evaluation is necessary to safeguard the humanitarian space and the perception of aid as independent and neutral. Coordination strategies range from coexistence (focusing on de-confliction minimising inconsistency, which is typical of conflict settings with military actors involved in active combat) to cooperation (focusing on harmonising a combined effort). CMCoord is essential to maintain a clear distinction between civil/humanitarian responders and the military, while avoiding duplication of efforts (Fig. 2.3).

United Nations Civil-Military Coordination (UN-CMCoord) formalises this essential dialogue and interaction to protect and promote humanitarian principles, avoid competition, minimise inconsistencies and, when appropriate, pursue common objectives. Good coordination requires information sharing and a clear sense of division of labour. UN-CMCoord can also help reduce the risk of misunderstandings or conflicts between military and civilian personnel and to ensure that humanitarian assistance is provided in a coordinated manner and consistent with International Humanitarian Law, where it applies.

Civ-Mil humanitarian coordination supports UNOCHA's overall efforts in humanitarian operations with a military presence. OCHA leads the creation and management of the interaction with military actors and supports humanitarian and military actors through training, advocacy on guidelines governing the use of foreign military, civil defense assets, and humanitarian civil-military interaction. The Civ-Mil relationship changes depending on the type of emergency and the roles and responsibilities of the military: it is therefore essential to define the reference context, the country in which one operates, the type of crisis being addressed, and the applicable legal framework. These aspects qualify the relations between civilians and military and define the contours of their relationships. The Last Resort principle always applies: a United Nations agency, through the Humanitarian Coordinator or the Government of an affected state, may submit a request for Foreign Military

[65] Bollettino and Anders (2020, pp. 3–10).

[66] CARE Emergency Toolkit (2022).

[67] Civil-Military Cooperation in Emergencies, European Commission, https://civil-protection-humanitarian-aid.ec.europa.eu/partnerships/relations/civil-military-cooperation-emergencies_en

Fig. 2.3 Civil-military cooperation in emergencies, European Commission. *Source* https://civil-protection-humanitarian-aid.ec.europa.eu/partnerships/relations/civil-military-cooperation-emergencies_en

and Civil Defense Assets (MCDA). The Civil-Military Coordination Service (CMCS) channels those requests through its network of government asset providers.[68]

The Comprehensive Approach[69] (CA), a consolidated concept within NATO for almost 20 years,[70] emerged from the lessons learned during the 1990s when the number and size of the international community's missions and interventions increased and NATO recognised that the military cannot resolve crisis or conflict by itself and that modern crises

[68] UNOCHA explanatory note 'Foreign Military and Civil Defence Assets in support to humanitarian emergency operations: what is the last resort?', https://www.unocha.org/publications/report/world/foreign-military-and-civil-defence-assets-support-humanitarian-emergency-operations-what-last-resort.

[69] NATO Standard AJP-3.16 Allied Joint Doctrine for Security Force Assistance (SFA) Edition A, Version, 1 May 2016, https://www.esercito.difesa.it/en/organization/The-Chief-of-General-Staff-of-the-Army/Training-Specialization-and-Doctrine-Command/Infantry-School/Documents/Doctrine-SFA/AJP-3-16-A-Allied-Joint-Doctrine-for-Security-Force-Assistance-SFA-Ed-maggio2016.pdf; Allied Joint Publication (AJP-01) Allied Joint Doctrine, Chap. 2; PO(2011)0045, Updated List of Tasks for the Implementation of the Comprehensive Approach Action Plan and the Lisbon Summit Decisions on the Comprehensive Approach, 04 March 2011.

[70] Riga Summit Declaration, https://www.nato.int/cps/eu/natohq/official_texts_37920.htm; Bucharest Summit Declaration, https://www.nato.int/cps/en/natolive/official_texts_8443.htm.

and conflicts were often not related to the military and therefore require assistance from outside the military.[71] The concept was slowly adopted not only by NATO but by the European Union and the United Nations as well. It is defined as a process aimed at facilitating system-wide coherence across the security, governance, development, and political dimensions of international peace and stability operations. Some differences remain however: the UN approach is concerned with the security-development nexus (peace, security, development, humanitarian elements). The EU approach focuses on the civil-military relationship (crisis management), while NATO's CA addresses its relationship with actors present on the ground of its operational mission, especially civilian organisations.

Agencies and actors can compete, coexist, coordinate, cooperate, integrate, or be united at the operational level. Humanitarian relief organisations are a set of actors which operate under a mandate that prioritises the humanitarian principles: they have no primary wish to be coherent with the political and military actors. Blurring these identities undermines the operational framework of the humanitarian community. This is why Civ-Mil relationships must be built on the recognition of each other's different roles. For humanitarian actors, the humanitarian principles are the overriding guide and coordination is something pragmatic, while for political and military actors, coherence and coordination are core operational principles.[72]

The levels of interaction between the Civ-Mil components are variable and the NATO CIMIC doctrine[73] (Civil-Military Cooperation) has established criteria that the military component must take into account. In addition to NATO's CIMIC doctrine,[74] which '*aims to enhance NATO military contributions to crisis response by improving NATO's military ability to interact with non-military actors*',[75] other tools followed such as the CMCOORD

[71] Wendling (2010).

[72] De Coning and Friis (2011).

[73] CIMIC Handbook: '*Understand non-military actors and respect their autonomy in decision-making and so encouraging them to do the same. Nevertheless, aim at promoting cooperation, reciprocal information sharing and unity of purpose if circumstances allow. Engage, via proactive communication and on respective level, with all non-military actors involved in the operation. Interact with respect, knowledge of respective roles, trust and transparency and be aware of cultural aspects. Incorporate non-military expert advice and factors, if applicable. Promote local ownership and build local capacity. Prepare for transition of non-military tasks as early as possible. Ensure internal NATO military coherence and consistent NATO messaging in interacting with non-military actors*'.

[74] Civil-Military Cooperation Centre of Excellence, https://www.cimic-coe.org/branches/cic/nato-cimic/.

[75] UNOCHA, Civil-Military Coordination; https://www.unocha.org/civil-military-coordination: '*Humanitarian organizations, State security forces and non-State armed groups increasingly operate in the same space in emergencies, whether they are caused by conflict, extreme weather events or other human-made disasters. Effective dialogue between these entities is critical to obtain and maintain humanitarian access to people in need and ensure their protection. UN Civil-Military Coordination (UN-CMCoord) formalizes this essential dialogue and interaction to protect and*

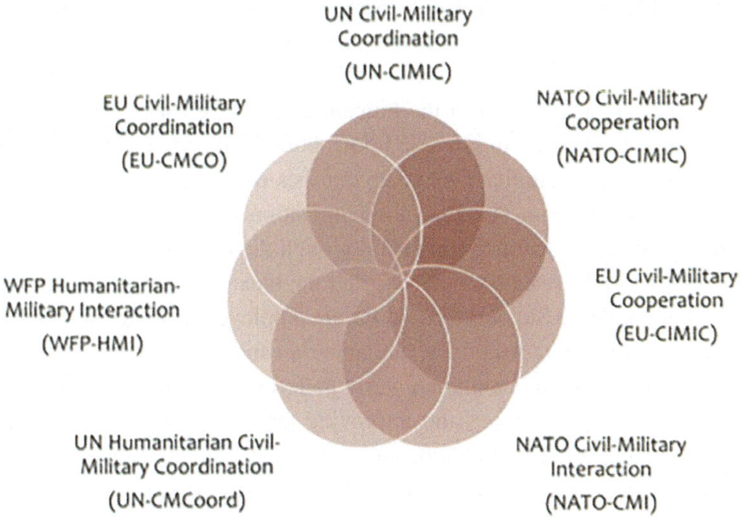

Fig. 2.4 Different civil-military concepts. *Source* Garbino et al. (2024, p. 9)

of the United Nations Organisation, the *Military and Civil Defence Assets (MCDA) guidelines for Complex emergencies,*[76] and the *Oslo Guidelines for Natural Disasters (MCDA for Natural Disasters)*[77] with the aim of promoting a more orderly interaction (Fig. 2.4).[78]

There are two insurmountable lines of demarcation for the two parties: the need for security and, at times, secrecy, of military information on one hand (in particular to preserve the '*Intelligence*' function), and the need to respect the humanitarian space on the

promote humanitarian principles, avoid competition, minimize inconsistency and, when appropriate, pursue common goals'.

[76] 'Guidelines On The Use of Military and Civil Defence Assets To Support United Nations Humanitarian Activities in Complex Emergencies', known as MCDA, Revision I, January 2006 significantly updated in 2018 by 'Recommended Practices for Effective Humanitarian Civil-Military Coordination of Foreign Military Assets (FMA) in Natural and Man-Made Disasters', Version 1.0 September 2018, https://www.unocha.org/publications/report/world/recommended-practices-effective-humanitarian-civil-military-coordination-foreign-military-assets-fma-natural-and-man-made-disasters-version-10-september-2018.

[77] See https://reliefweb.int/report/world/oslo-guidelines-guidelines-use-foreign-military-and-civil-defence-assets-disaster-relief-revision-11-november-2007.

[78] For a quick overview see: https://emergency.unhcr.org/sites/default/files/OCHA%2C%20On%20Message%20%E2%80%93%20Civil%20Military%20Coordination%2C%20November%202013.pdf. The list of the existing tools include: (i) Guidance Note on OCHA CMCoord Support to Protection Outcomes (2020); (ii) Guidelines on the Use of Military and Civil Defence Assets in Disaster Relief ('Oslo Guidelines') (2007); (iii) IASC Non-Binding Guidelines on the Use of Armed Escorts for Humanitarian Convoys (2013); (iv) IASC Guideline: Civil-Military Relationship in Complex Emergencies; (v) Recommended Practices for Effective Humanitarian Civil-Military Coordination of Foreign Military Assets in Natural and Man-Made Disasters (2018).

other. The blurring of lines, even if only perceived, visual or cognitive by the population on whose behalf civilian and military actors intervene, would be harmful for all the parties involved.

Currently there is a reduction in the humanitarian space for civil organisations, i.e. a lesser possibility of operating in safe conditions both for the victims and for the personnel providing assistance. The military mandate consists of putting into practice the political will of those who decide to start the relief operation. Once governments have defined the objectives, these are identified militarily through the mandate, which establishes security on the ground, the strategic objectives, and the means available to achieve them. The military working in Peace Support Operations are employed for functions such as observation, liaison, occasionally protection of humanitarian and refugee convoys, support to civil agencies and medical and humanitarian work, therefore they must have greater adaptability. This includes also developing and improving the space for interaction with civilians in humanitarian support, coordination, and consensus-building activities. Humanitarian workers should never present themselves and their work as part of a military operation, nor should military personnel present themselves as humanitarian personnel.[79] The '*do no harm*'[80] commitment and the Last Resort principle must guide any coordination level.

2.3 Coordination and Relief Standardisation

Coordination depends substantially on sharing a common language and following the same standards. The language, operational terminology, and standards must be aligned to ensure coherence and interoperability. The multiple attempts to achieve this, and the ongoing discussions among many different constituencies to introduce standards for similar activities, are testament to the imperative to realise this objective.

2.3.1 INSARAG

The INSARAG is an international network of Search And Rescue professionals that provide guidelines and technical standards for Search and Rescue activities. INSARAG was established in 1991 in the aftermath of the response to the 1988 Armenian earthquake, during which the difference of standards and procedures among different SAR teams were clearly evident. Based on their Guidelines, INSARAG also foresees an internationally recognised classification for Urban Search And Rescue Teams. The OSOCC concept was originally developed in the INSARAG guidelines as a tool for the coordination on

[79] Osservatorio di Politica Internazionale (ISPI, IAI, CeSI, CeSPI), Rapporto collettivo - Le missioni internazionali (2010, pp. 102–143), https://www.parlamento.it/documenti/repository/affariinternazionali/osservatorio/rapporti/RA2010_collettivo.pdf.

[80] Anderson (1999).

SAR. A dedicated classification process has been established to verify the quality of the national teams.

2.3.2 Emergency Medical Teams (EMT)

Following the example of INSARAG and the lessons identified after the earthquake of Haiti of 2010, the WHO took the lead of the EMT initiative to set up internationally recognised standards for emergency medical teams to be employed in case of disaster. After a long negotiation among experts, three level of standards were identified in the 'Blue book' in 2010. The standards were updated later to include other medical capacities. WHO has also defined a specific coordination setting for EMTs, the EMTCC which will work in connection with UNDAC teams when deployed and in strong cooperation with the Ministry of Health of the affected state.

2.3.3 EU Civil Protection Modules[81]

In the framework of the European Civil Protection Mechanism, the concept of civil protection modules was developed in the aftermath of the response to the 2004 South East Asia Tsunami. The European Commission and Member States have developed their standards and rules of engagement: 29 different typologies of civil protection modules have been identified.[82] Tasks, capacities, main components, and deployment times are defined and standardised and a certification process has been established for all interested teams.

2.3.4 The Sphere Project

Following the example of the IASC, a number of inter-agency initiatives were developed aimed at improving accountability, quality, and performance of humanitarian action. Hundreds of agencies, bilateral and multilateral donors, the UN system, the Red Cross and Red Crescent Movement, NGOs, and inter-agency umbrella organisations have participated in, or are members of, one or more of these initiatives. The Sphere Project[83] is probably the most well-known and implemented. Launched in 1997 by a group of humanitarian NGOs and the Red Cross and Red Crescent Movement, Sphere has developed a

[81] For more details, see Billing's chapter in the present volume.

[82] Commission Implementing Decision (EU) 2025/704 of 10 April 2025 laying down rules for the implementation of Decision No 1313/2013/EU of the European Parliament and of the Council on a Union Civil Protection Mechanism and repealing Commission Implementing Decisions 2014/762/EU and (EU) 2019/1310.

[83] Sphere (2018).

handbook of standards for four sectors (Water/Sanitation and Hygiene Promotion; Food Security; Nutrition and Food Aid; Settlement and non-Food Items and Health Services). Each standard has key indicators and guidance notes.

Among the various initiatives that were developed it is worth to mention: the Active Learning Network for Accountability and Performance in Humanitarian Action (ALNAP).[84] The International Organisation for Standards (ISO) is also developing standards in response to risks of natural and anthropogenic origin. These standards are set to include guidelines on 'command, control, coordination and cooperation' in disaster response, in areas such as 'information gathering; information sharing/information processing; information flows/knowledge flow; interoperability; structures and procedures; decision support; and warning'.[85]

2.4 Conclusions

The field of emergency and crisis management at the international level is becoming increasingly crowded and complex. In a historical phase in which the geopolitical relations that have governed the past decades are being questioned, attempts to reconsider the traditional actors and the very principles that have governed relations between different countries are being witnessed, as well as humanitarian action. In this context, an increase in needs on a global scale is observed with a consequent competition for available financial resources that is exacerbated by the substantial change in the nature and number of actors from different geographical, cultural, and operational backgrounds.

The fundamental challenge is to understand how to ensure the relevant actors work together to avoid gaps, overlaps, and duplications. Coherence should be ensured between the strategic and the operational levels, and among international, national, and local levels. To achieve this, it is crucial to invest in preparedness at all different levels.

Starting from the local level, a dedicated clear legal and operational framework must be defined on the basis of internationally recognised standards. All countries should be able to implement their sovereign role in managing the emergencies on their territory, while at the same time an additional effort should be made to implement the commitments of *The Grand Bargain* related to the involvement of affected communities in the decision-making. At all levels different entities should have available dedicated personnel, trained and aware

[84] For more information, Advancing humanitarian learning, https://alnap.org/. ALNAP is a network that incorporates many of the key humanitarian organizations and experts from across the sector, including members from donor, NGO, Red Cross/Crescent, UN and independent/academic organizations. ALNAP aims at improving humanitarian performance through learning and accountability and consists of core activities and projects. ALNAP access to a broad range of experience within its membership to produce tools and analyses relevant and accessible to the humanitarian sector as a whole.

[85] International Organisation for Standardization, Business Plan-ISO/TC 292 Security and resilience. https://www.iso.org/committee/5259148.html.

of international principles and dynamics. Coordination should not become the problem but the solution.[86] At international level an additional effort should be made to rationalise the existing coordination tools as well as to standardise means and procedures. The number of initiatives and constituencies addressing similar concerns raises an additional coordination problem. There is an emerging need of having verifiable and known standards with a clear set of evaluation indicators.

Rescuers should be accountable first of all to their stakeholders, especially the affected population and institutions, and the international community must make an additional effort in identifying the organisations and entities that have the best capacities to cope with the different post-disaster phases. This must take place across all strata: donors, UN, and NGO/CSOs. Both at national and international levels, greater emphasis should be given to the needs assessment to avoid gaps and overlaps in the response. The urgency of providing humanitarian relief must be balanced with the necessity of providing needs-based resources. The issue of civil-military coordination raises specific concerns. In complex emergencies, the clear difference of roles and mandates should be maintained, applying the *last resort* and the *do no harm* principles. The best way to avoid conflicting interests between humanitarian and political/strategic concerns is to reiterate and reinforce the value of the humanitarian principles. Organisations unable or unwilling to respect the standards should be excluded, even though this is likely to be extremely controversial.

Key Sentences

- **The Growing Complexity of Crisis Response Action**—The number of actors active in emergency and crisis management has increased significantly in recent years: the coexistence of large complex organisations in the same operational environments increases the level of complexity, with sometimes poorly coordinated efforts on a common set of objectives.
- **The Primary Responsibility of the Affected State**—Each country is responsible for managing the consequences of events on its own territory. It must define a legal framework and have trained structures and personnel; the country can also request international assistance to coordinate it when on its territory.
- **An Established International Humanitarian Assistance Framework**—The international community has organised itself to provide assistance to countries that request it. Based on the fundamental principles of humanitarian action (humanity, neutrality, impartiality, and independence) and the concept of do no harm, tools, common languages, guidelines, and standard operating procedures have been defined to improve the efficiency and speed of intervention for both civilian and military actors.

[86] Boin and Bynander (2014).

- **Fostering Civil-Military Coordination**—Civilian and military actors operate under different mandates, strategic interests, principles, roles, doctrines, value systems, and operational modalities, sharing a temporal space that requires communication, interaction, and coordination. An effective dialogue among these entities is essential to gain and maintain humanitarian access to the people in need, clarifying the lines of command and the division of specialised work, respecting the humanitarian space and applying the last resort principle. EU, NATO, and the UN all have their own guidelines and tools to regulate the civ-mil interaction.
- **Standardised Procedures and Coordination Tools**—International response teams and coordination structures have developed highly detailed, specialised procedures and tools that serve as fundamental reference points. It is important that all stakeholders involved follow these procedures and use those tools.
- **The Need for Context-Specific Coordination and Responsibility Allocation**—Depending on the country affected by the emergency or crisis and the type of event, the various tools and their reference framework vary in a flexible manner. It is necessary to have a clear understanding of how coordination and the division of responsibilities work in each different emergency.

References

Adinolfi C et al (2005) Humanitarian response review, an independent report commissioned by the United Nations Emergency Relief Coordinator & Under-Secretary-General for Humanitarian Affairs, Office for the Coordination of Humanitarian Affairs

Alejandria MCP et al (2022) Humanitarian-military relations in complex emergencies: evidence, insights, and recommendations. Report of the Center for Human Rights and Humanitarian Studies at the Watson Institute for International Public Affairs. Brown University

Alexander D (2002) From civil defence to civil protection—and back again. Disaster Prev Manag 11:209–213

ALNAP (2010) The role of national governments in international humanitarian response to disasters. In: Meeting Paper 26th annual meeting, 16–17 November 2010, Kuala Lumpur, Malaysia. https://alnap.org/help-library/resources/the-role-of-national-governments-in-international-humanitarian-response-to-disasters/

ALNAP (2022) The state of the humanitarian system (SOHS). https://alnap.org/help-library/resources/2022-the-state-of-the-humanitarian-system-sohs%2D%2Dfull-report/

Anderson MB (1999) Do no harm: how aid can support peace-or war. Lynne Rienner Publishers, Boulder

Bell SR et al (2019) Help or hindrance? The role of humanitarian military interventions in human security NGO operations. Int Political Sci Rev 40:263–278

Bennett J et al (2006) Coordination of international humanitarian assistance in tsunami-affected countries. Tsunami Evaluation Coalition

Boin A, Bynander F (2014) Explaining success and failure in crisis coordination. Geogr Ann Ser A Phys Geogr 97:123–135

Bollettino V, Anders B (2020) Civil-military coordination: a framework for measuring effectiveness in humanitarian response. J Hum Affairs 2:3–10

CARE Emergency toolkit (2022). https://www.careemergencytoolkit.org/topics-issues/11-interacting-with-armed-actors/

De Coning C, Friis K (2011) Coherence and coordination. The limits of the comprehensive approach. J Int Peacekeep 1:243–272

De Siervo G (2012) Actors, activities and coordination in emergencies. In: de Guttry A et al (eds) International disaster response law. Springer, pp 485–516

Development Initiatives, DI (2024) Falling short? Humanitarian funding and reform. https://devinit.org/resources/falling-short-humanitarian-funding-reform/#exec-summary

Donini A et al (2008) The state of the humanitarian enterprise. Tufts University, Feinstein International Center

Garbino H et al (2024) Civil-military what?! Making sense of conflicting civil-military. Center for Human Rights and Humanitarian Studies. Watson Institute for International and Public Affairs, Brown University and Department of War Studies, Swedish Defence University

Global Humanitarian Platform (2010) Strengthening partnership for effective humanitarian action, 2–3 February, 2010, Summary note, p 2. https://www.icvanetwork.org/uploads/2021/09/GHP-Summary-Note-Feb-2010.pdf

Grace R et al (2023) Moving humanitarian-military relations forward: a new typology. Int J Humanitarian Action 8:2

Grünewald F, Binder A (2010) Inter-agency real-time evaluation in Haiti: 3 months after the earthquake. Groupe URD and the Global Public Policy Institute

Hofman M (2014) The evolution from integrated missions to 'peace keepers on steroids': how aid by force erodes humanitarian access. Glob Responsibil Protect 6:246–263

Humanitarian Action, Global Humanitarian Overview, (2025), Trends in crises: a world on fire (4 December 2024). https://humanitarianaction.info/document/global-humanitarian-overview-2025#page-title

IASC (1994) Working group XVI meeting, definition of complex emergencies, p 2

IFRC (2007) Annotations to the draft guidelines for the domestic facilitation and regulation of international disaster relief and initial recovery assistance

IFRC (2008) Introduction to the guidelines for the domestic facilitation and regulation of international disaster relief and initial recovery assistance

Inter-Agency Standing Committee, IASC (2004) Civil-military relationship in complex emergencies, an IASC reference paper. https://interagencystandingcommittee.org/sites/default/files/migrated/2021-03/Civil-Military%20Relationship%20in%20Complex%20Emergencies.pdf

ICVA (2024) Adapting humanitarian coordination challenges, innovations, and future directions for NGO-led mechanisms. https://www.icvanetwork.org/uploads/2024/10/Adapting-Humanitarian-Coordination.pdf

Katoch A (2006) The responders' cauldron: the uniqueness of international disaster response. J Int Aff 59:153–172

Kent R (2011) Narrowing the corridors of humanitarian action, humanitarian futures programme. King's College London, UK, pp 2–3

Lilly D, Bowden M (2024) A shrinking humanitarian marketplace. Time for better regulation. Centre for Humanitarian Action. https://www.chaberlin.org/wp-content/uploads/dlm_uploads/2025/01/250123-humanitarian-marketplace-web-2.pdf

Macrae J (ed) (2002) The new humanitarianisms: a review of trends in global humanitarian action. Humanitarian Policy Group Report 11. Overseas Development Institute

Müller-Stewens G et al (2019) The professionalization of humanitarian organizations. Springer
Niinistö S (2024) Safer together: A path towards a fully prepared union—Strengthening Europe's civil and military preparedness and readiness: Report by Sauli Niinistö, former President of the Republic of Finland in his capacity as Special Adviser to the President of the European Commission. https://commission.europa.eu/topics/defence/safer-together-path-towards-fully-prepared-union_en
OECD (2010) 'Review of the Italian National Civil Protection System. OECD international futures Programme', pp 18 and 44
OSOCC(2018) OSOCC guidelines OCHA. https://insarag.org/wp-content/uploads/2020/04/OSOCC_Guidelines_2018_English_2.pdf
Pearson M (2024) New money? What the numbers say about 'non-traditional' aid donors. The New Humanitarian, Aid and Policy Analysis. https://www.thenewhumanitarian.org/analysis/2024/04/30/new-money-what-numbers-say-about-non-traditional-aid-donors
Polman L (2010) The crisis caravan: what's wrong with humanitarian aid? Metropolitan Books
Sphere (2018) Humanitarian charter and minimum standards in disaster response, the Sphere project, https://www.spherestandards.org/handbook/
The Tripartite Core Group (2008) Post-Nargis joint assessment, p 46. https://www.gfdrr.org/sites/default/files/GFDRR_Myanmar_Post-Nargis_Joint_Assessment_2008_EN.pdf
Varga C (2020) Civil society associations versus so-called non-governmental organisations. Polgári szemle 16:212–225
UNDAC (2024) UNDAC handbook version 2, 8th edn. https://undac.un.org/en/article/undac-handbook-8th-edition-version-2-june-2024
UNGA Res 57/150 (2002) Strengthening the effectiveness and coordination of international urban search and rescue assistance
Wendling C (2010) The comprehensive approach to civil-Miltary crisis management—A critical analysis and perspective. IRSEM Report

Open Access This chapter is licensed under the terms of the Creative Commons Attribution 4.0 International License (http://creativecommons.org/licenses/by/4.0/), which permits use, sharing, adaptation, distribution and reproduction in any medium or format, as long as you give appropriate credit to the original author(s) and the source, provide a link to the Creative Commons license and indicate if changes were made.

The images or other third party material in this chapter are included in the chapter's Creative Commons license, unless indicated otherwise in a credit line to the material. If material is not included in the chapter's Creative Commons license and your intended use is not permitted by statutory regulation or exceeds the permitted use, you will need to obtain permission directly from the copyright holder.

Civil Protection and Disaster Management in the European Union: Actors, Structures, and Processes

3

Peter Billing

Abstract

This chapter analyses the role of the European Union in managing disasters and disaster risks. Following a short description of the predominant disaster risks and the national disaster management structures in Europe, the key instruments for disaster risk management at European level are presented. Primary attention is given to the evolution, mandate, instruments, and operational practice of the Union Civil Protection Mechanism (UCPM). The UCPM has become the most sophisticated disaster management instrument at European level. Its *modus operandi* and interaction with other instruments are explained and illustrated through numerous examples. The final section provides a short outlook of trends, challenges, and opportunities the UCPM is facing in a changing international risk landscape.

Keywords

European Union • Union Civil Protection Mechanism (UCPM) • Emergency Response Coordination Centre (ERCC)

3.1 Introduction

The present chapter analyses the role of the European Union (EU) in managing disasters and disaster risks. It will provide insights into the different facets of the Union Civil Protection Mechanism (UCPM, also referred to as the 'Mechanism'). The Mechanism is a

P. Billing (✉)
Department of Geography, University of Bonn, Bonn, Germany
e-mail: pbillng@uni-bonn.de

© The Author(s) 2026

E. Sommario and P. Billing (eds.), *International Legal and Institutional Frameworks for Disaster Risk Management*, Civil Protection, Disaster Management, and Resilience 4,
https://doi.org/10.1007/978-3-032-11806-6_3

collaboration arrangement of disaster management authorities from 37 European countries and the European Commission.[1] The UCPM has become a key player in managing natural and man-made disasters not only in Europe, but also globally. Its evolution, main actors, structures, processes, and instruments, as well as the way lessons from operational practice have shaped its current architecture, thus deserve a closer look.

Before delving deeper into the architecture of the Mechanism, we will briefly describe the predominant hazards, risks and disasters Europe faces (Sect. 3.2). Section 3.3 then provides a brief typology of the civil protection structures in EU Member States and Participating States. It describes the fundamentals of the European disaster management system at national level and its interaction with the EU. Section 3.4 presents an overview of the evolution and main components of the UCPM: objectives, scope and definitions, mandate and main instruments available at different stages of the disaster cycle (prevention, preparedness, response). Selected examples from operational practice will illustrate the *modus operandi* of the UCPM, how its various instruments interact and which challenges it faces. A brief overview of other key EU instruments in the area of risk management, in particular, regarding flood management, forestry, chemical accidents, nuclear safety, and marine pollution, will complement the picture.

Taking stock of the Mechanism's accumulated experience, Sect. 3.5 finally provides a short outlook of the trends, challenges, and opportunities for the UCPM to maintain and even further improve the level of protection for citizens in Europe and abroad in a changing risk landscape.

[1] The Mechanism is composed of the 27 EU Member States plus 10 so-called 'Participating States': Albania, Bosnia and Herzegovina, Iceland, Moldova, Montenegro, North Macedonia, Norway, Serbia, Türkiye, Ukraine. The 10 Participating States contribute to the budget and can participate in all activities of the Mechanism. However, they do not participate in formal votes. For ease of readability these are subsumed under 'Member States' except when we touch on specificities that concern only these states in their role as Participating States.

3.2 Hazards, Risks, and Major Disasters

Disasters have spread death and destruction across populations in Europe throughout history.[2] A long list of catastrophic events can be drawn, stretching from natural disasters like the eruption of the Santorini Volcano in ancient times to our modern era, when large-scale industrial accidents like the release of toxic substances from the Seveso plant in 1976 added anthropogenic ('man-made' or 'human-induced') disasters to that list.

Apart from their direct impact in terms of fatalities, injuries, or economic damage as outlined in the introduction of this volume, major disasters can also have massive political repercussions on countries. They can even have the power to fundamentally alter entire socio-political systems. Following the 1755 earthquake in Portugal, for instance, the Marquis de Pombal rebuilt and reformed the country's state institutions, thus laying the foundations of modern Portugal, creating the basics of modern seismology, and unleashing 'an intellectual earthquake in Europe of the Enlightenment'.[3] Even in modern times, the way governments manage disasters can have a decisive impact on their political fate. As disasters know no borders, it does not require a Tambora-type volcanic eruption, as in 1815, to understand that even geographically confined incidents can have large-scale impacts beyond national borders and even continents.[4]

3.2.1 Data and Analytical Tools

Over the last few decades, scientists have undertaken significant efforts to collect data and develop models to identify, localise, and assess hazards, risks, and disasters. Datasets with global coverage were developed by several organisations, such as Munich Re's Nat-Cat (Natural Catastrophe) dataset, the World Bank's research on global natural disaster hotspots, the United Nations Environmental Programme (UNEP) Disaster Risk Index (DRI) or the Centre for Research on the Epidemiology of Disaster's (CRED) EM-DAT disaster database. Extensive efforts on global risk analysis are also being undertaken by

---

[2] Internationally, the most widely used definition of 'disaster' is presented by the United Nations Office for Disaster Risk Reduction (UNDRR): 'a serious disruption of the functioning of a community or a society at any scale due to hazardous events interacting with conditions of exposure, vulnerability and capacity, leading to one or more of the following: human, material, economic and environmental losses and impacts' UNDRR (2017). For statistical purposes, some institutions like the Centre for Research on the Epidemiology of Disasters (*CRED*) (https://www.emdat.be/) provide quantitative elements as well. A broader discussion of concepts and terminology related to hazards, risks and disasters is provided in the introductory chapter of this volume.

[3] Fonseca (2005, p. 106).

[4] Wood (2014).

a consortium led by the European Commission through its Index for Risk Management (INFORM) and Welle and Birkmann's World Risk Index (WRI).[5]

While the above projects focus on the country level and global coverage, data from other projects allow analyses at a finer scale: the European Spatial Planning Observation Network (ESPON), for example, has been providing hazard analyses at regional level for Europe since 2002 and continuously improved its data and methodology.[6] It provides a comprehensive overview of the geographical distribution of specific natural (hydro-meteorological, geophysical) and some technological hazards (chemical accident, oil spill, nuclear accident) across Europe.

The above analytical tools provide useful helicopter views of the spatial distribution of hazards at continental, country, or regional scale, but they are of limited usefulness for local hazard assessments. For analyses requiring a fine grading, the EU's Copernicus programme is a useful instrument as it can provide high-resolution hazard or risk maps for local areas.[7]

EU Member States are required to establish national risk assessments and risk management plans. A summary of those plans is shared with the European Commission, which publishes aggregated reports of these assessments. By capturing key trends in Europe's disaster risk landscape and its main drivers, the reports facilitate evidence-based decision-making on risk management. As can be seen in Fig. 3.1, the risk most frequently mentioned in the latest report is flooding, followed by nuclear accidents, extreme weather, epidemics, industrial risks, wildfires, and drought.[8]

3.2.2 Major Disasters

As already stated in the introduction of this volume, the number of natural disasters has significantly increased globally since the beginning of the twentieth century, with 350–450 disasters every year, leaving behind a trail of death and destruction, millions of homeless,

[5] See Munich Re (2025), Dilley et al. (2005), Peduzzi et al. (2009), and the CRED EM-DAT disaster database, available at https://www.emdat.be, last accessed (as any subsequent URL) on 03 June 2026; Welle and Birkmann (2015), Bündnis Entwicklung Hilft (2025), European Commission, Joint Research Centre (2024). For a critical review of the respective methodological approaches see Mazhin et al. (2021).

[6] The geographical distribution of the hazards used by ESPON (https://www.espon.eu) is the NUTS-3 level. NUTS is a classification developed by the EU to reference countries' regions for statistical purposes (Nomenclature of Territorial Units for Statistics). NUTS divides each EU country into 3 levels: major socio-economic regions, basic regions, and small regions (NUTS-3). See also Klein et al. (2024), Schmidt-Thomé (2006a), Schmidt-Thomé (2006b), Cantergiani et al. (2021), Navarro et al. (2023).

[7] Cf. Copernicus (2025). https://mapping.emergency.copernicus.eu.

[8] European Commission (2024a, p. 7).

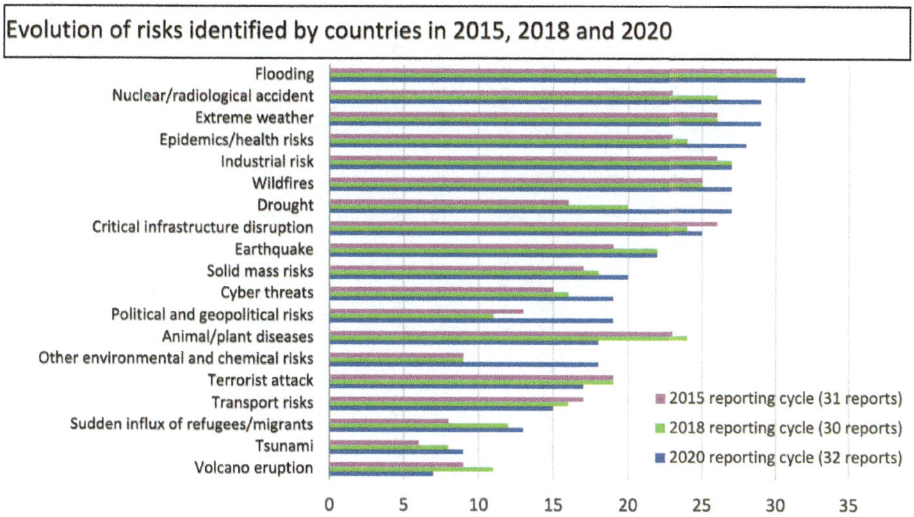

Fig. 3.1 Number of countries identifying a specific disaster risk as relevant in 2015, 2018, and 2020. *Source* European Commission (2024a, p. 7)

and huge economic damages.[9] Even though overall fatalities have decreased due to better preparedness compared to the early twentieth century, new disaster risks are emerging, compounding natural, technological, and biological risks.

Within Europe, the human impact of disasters is equally significant, albeit less pronounced than in Africa or Asia. The most destructive disasters in Europe in terms of fatalities, people affected, and economic damage between 1945 and 2025 are presented in Table 3.1.

From these figures, one can draw four conclusions: (1) heatwaves have become the dominant natural disaster type causing fatalities in Europe; (2) floods and, to some extent, earthquakes cause the biggest damage for the economy; (3) droughts and floods affect the largest number of people; 4) technological disasters are generally less damaging in Europe.

3.2.3 Disaster Trends

Most experts agree that climate change is already impacting weather and climate extremes worldwide and has led to increasingly negative impacts on nature and people alike.[10] Heatwaves, for instance, have received increasing attention in recent years due to global

[9] Ritchie et al. (2022). See also CRED (2020) and CRED (2023).

[10] Intergovernmental Panel on Climate Change, IPCC (2023, p. 5). The European Environment Agency (EEA) estimates economic damage in Europe caused by climate change amounting to 822

Table 3.1 Most destructive disasters in UCPM member states and participating states

Fatalities				Affected persons				Economic damage			
Disaster type	Country	Year	Total deaths	Disaster type	Country	Year	Total affected	Disaster Type	Country	Year	Damage ('000 USD)
Heat wave	Italy	2003	20,089	Drought	Spain	1990	6,000,000	Flood	Germany	2021	40,000,000
Heat wave	France	2003	19,490	Extra-tropical storm	France	1999	3,400,011	Earthquake	Italy	1980	20,000,000
Heat wave	Italy	2022	18,010	Drought	Albania	1989	3,200,000	Earthquake	Italy	2012	15,800,000
Heat wave	Spain	2003	15,090	Blizzard/ winter storm	Moldova	2000	2,600,000	Flood	Germany	2013	12,900,000
Heat wave	Italy	2023	12,743	Riverine flood	Ukraine	1995	1,700,000	Flood	Germany	2002	11,600,000
Heat wave	Spain	2022	11,324	Flood (general)	Slovenia	2023	1,500,120	Flood	Spain	2024	11,000,000
Heat wave	Germany	2003	9355	Riverine flood	Italy	1970	1,301,650	Industrial accident	Spain	2002	9,960,407
Heat wave	Spain	2023	8352	Flood (general)	Italy	1966	1,300,000	Flood	Italy	2023	9,750,000
Heat wave	Germany	2022	8173	Riverine flood	Czechia	2013	1,300,000	Flood	Italy	1994	9,300,000
Heat wave	Germany	2023	6376	Flood (general)	Romania	1975	1,000,000	Storm	France	1999	8,000,000

Source EM-DAT (author's own calculations from EM-DAT database, 20/02/2025)

warming. Scientific studies estimate that more than 70,000 people have been killed in Europe by the heatwave of 2003, and more than 60,000 during the summer in 2022.[11] According to the European Environment Agency (EEA), heat-related hazards are also increasing in the northern hemisphere, such as in Germany or Scandinavia.[12]

Global warming is likely to have knock-on effects on food security through agricultural insect pests and decreasing crop yields due to drought, as well as on public health due to the spread of vector-borne diseases.[13] Environmental, demographic, and geopolitical changes, as well as disruptive new technologies, further contribute to increasing the risks and pose new challenges.[14] Disaster-(risk) management, therefore, remains an area where constant improvements are necessary to keep pace with a dynamically evolving global disaster landscape.[15] One such area is the enhancement of disaster management structures at the national and international levels, which will be explored in the following section.

3.3 National Disaster Management Structures: Civil Protection Systems in Europe

The previous section has provided an overview of the wide spectrum of hazards, risks, and disasters European countries are facing. Some of them experience a high frequency and magnitude of different types of disasters. Italy, for instance, faces almost all types of high-impact disasters, encompassing floods, earthquakes, volcanoes, forest fires, and large-scale industrial accidents. Other countries, such as Malta or Finland, experience much less such high-impact disasters due to their somewhat more 'privileged' geographical location. Yet other countries are confronted by a few predominant types of hazards, such as the Netherlands as regards water-related hazards.

It is therefore not surprising that the national risk and disaster management systems developed by European countries vary, reflecting their risk profile, historical experience, their organisational and political principles. However, there are also common features and similarities to be noted, following some basic principles of disaster management. Scientific literature on disaster management generally identifies a number of characteristic elements common to all disaster management systems designed to ensure effective

billion EUR in the period 1980–2024. https://www.eea.europa.eu/en/analysis/indicators/economic-losses-from-climate-related.

[11] Ballester et al. (2023). See also World Meteorological Organisation (2025) Heatwave. https://wmo.int/topics/heatwave.

[12] EEA (2021) Forest Fires in Europe. https://www.eea.europa.eu/en/analysis/indicators/forest-fires-in-europe.

[13] Subedi et al. (2023), Brugueras et al. (2020), EEA (2022).

[14] Niinistö (2024, p. 32).

[15] UNDRR (2023).

organisation in different phases of the disaster cycle. The core elements will be presented below.

3.3.1 General Characteristics of Disaster Management Systems

Researchers have identified some core characteristics of disaster management systems that apply, with variations, anywhere in the world.[16] The three essential components encompass:

First: Existence of specific organisational structures, i.e. standardised command, control and coordination structures for emergency response involving one or more agencies at different levels. The scale of a disaster usually determines at which level it is managed, starting from the local level to one (or more) regional level(s) to the national level, or even involving international assistance.

Second: A comprehensive emergency management approach, which integrates all phases of disaster management into one cohesive, hazard-independent framework. Comprehensive emergency management encompasses structured and continuous measures in each segment of the disaster cycle. For *prevention,* these measures encompass three core elements: risk assessment, mitigation measures to minimise the adverse effects of disasters, and policy/legislative elements supporting disaster risk reduction efforts. In the *preparedness* segment, crucial elements are training, exercises, and education for emergency responders and the potentially affected community, the development of early warning systems, as well as resource management, i.e. availability and accessibility of necessary resources (medical supplies, equipment stockpiles, and emergency shelters). In the *response* phase, the characteristic elements are: rapid resource mobilisation to affected areas, clear and consistent communication channels among all parties involved, as well as immediate medical care and psychological support to affected individuals and responders.

Third: Existence of coordination and communication procedures to ensure smooth interaction among various stakeholders, including government agencies, Non-Governmental Organisations (NGO), and the population.

The bulk of scientific literature focuses on governments as the main actors in any disaster management system, as they bear the ultimate responsibility for disaster management.[17] In most countries, the key governmental players in disaster management are the civil protection authorities. They play a pivotal role in their respective national disaster management systems, no matter what their specific organisational model.[18] Within the UCPM, they are also the main interface to the European and international level. EU legislation assigns specific roles and obligations to them.

---

[16] Cf. Lettieri et al. (2009) and López-Carresi (2012).

[17] Lettieri et al. (2009, p. 128).

[18] It is acknowledged that private industry, volunteers, and the population equally play important roles, but due to space limitations this topic will not be covered in depth in this chapter.

3.3.2 National Civil Protection Systems in Europe: A Comparative Perspective

Comparing the national civil protection (CP) systems generates interesting research questions, in particular whether there are specific features, organisational principles, or procedures in these systems that are more conducive to facilitating successful disaster management than others. However, literature on national CP systems in Europe is fragmented and patchy, in contrast to what one might expect, given their crucial role in managing an increasing number of increasingly complex disasters.

The Directorate-General for European Civil Protection and Humanitarian Aid Operations (DG ECHO)'s synoptic overview of the national systems in Europe provides a useful first orientation.[19] Most of the research attention paid to European civil protection matters has so far been focused on its role within the EU's overall crisis management capacities.[20]

The number of studies with a detailed single-country focus or a systematic comparative focus on the CP systems of two or more countries is still limited.[21] The pioneering work of the ANVIL (Analysis of Civil Security Systems in Europe) project, covering 22 countries, is still the most comprehensive study offering a general orientation on the subject.[22] The following overview of national civil protection structures in Europe will therefore essentially draw on findings of this project, complemented, where necessary, by information provided by Member and Participating States through DG ECHO's website (https://civil-protection-humanitarian-aid.ec.europa.eu/index_en) and some studies with a more limited coverage. The comparison essentially covers five dimensions: (1) institutional set-up (parent authority, centralised/decentralised organisational model, role of non-state organisations); (2) designated role of military in domestic emergency management; (3) intervention thresholds; (4) procedures and arrangements; and 5) quality safeguarding mechanisms.

Concerning organisational set-up, the parent authority to which civil protection is assigned in most countries is the Ministry of the Interior. In two countries, however, it is assigned to the Ministry of Justice (Iceland, Netherlands); in four countries to the Ministry of Defence (Denmark, Ireland, Slovenia, Sweden); and in other countries either directly to the Prime Minister's Office (Italy), or to a specific Ministry dedicated to crisis management (Greece) (based on data from ECHO's overview above).

---

[19] ECHO (2025).

[20] Cf. Backman and Rhinard (2018), Boin et al. (2013), Morsut (2014), Villani (2017), Widmalm et al. (2019), Kalogiannidis et al. (2024).

[21] Spassov and Tagarev (2015), Zweglinski and Cordero (2019) provide single-country studies while Talay (2020), Morsut and Kruke (2020), Petersson (2023), Christensen et al. (2016), Prior et al. (2016), Esposito (2017), Dadoudi (2017), Laakso (2019), Brazova (2021) opted for a comparative focus on the CP systems of two or more countries.

[22] Boin et al. (2014), Kuipers et al. (2015).

The degree of centralisation of national CP systems—measured by way of criteria like delegation of decision-making powers, executive responsibility, and policy formulation—varies considerably (see Fig. 3.2). With the exception of France, the degree of centralisation seems to be higher in eastern European countries than in the rest of the continent.

Volunteer organisations and the private sector contribute to civil security in most countries, albeit to varying degrees. Brazova (2021) investigates the role of non-state actors in national disaster management. She identifies a cluster of countries with a very state-centric approach (Czechia, Poland, Ireland, United Kingdom, Malta) where the official role of non-state actors is relatively marginal. In contrast to this, in a cluster formed by Hungary, Italy, and Romania, the role of the non-state sector is rather important, both in regular civil security provision and during crisis situations. Similarly, both Germany and Austria are characterised by strong formal volunteerism of civil society and private organisations with legally defined roles in disaster management. Brazova concludes by stating that 'although the state plays a crucial role in civil security provision in all European

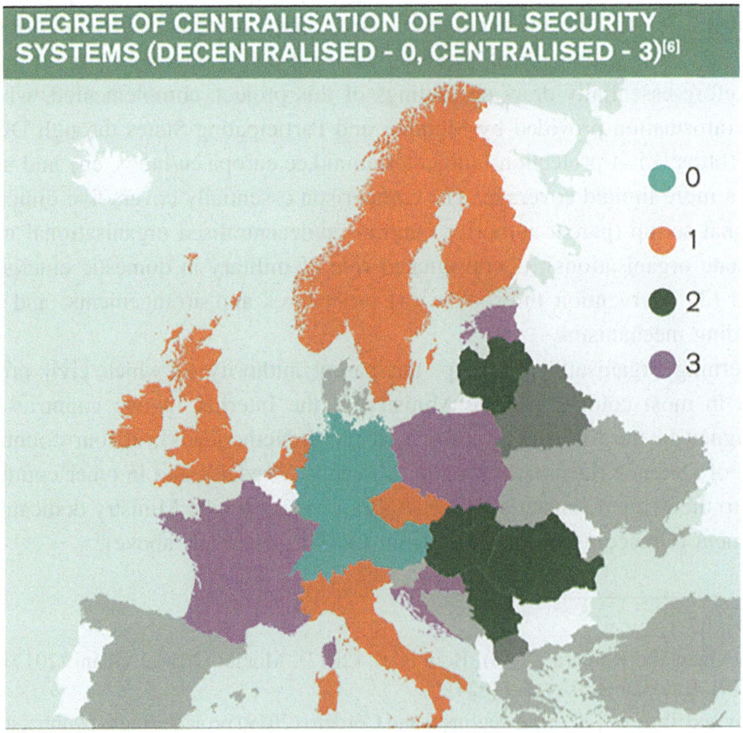

Fig. 3.2 Degree of centralisation of civil security systems in Europe. *Source* Boin et al. (2014, p. 7)

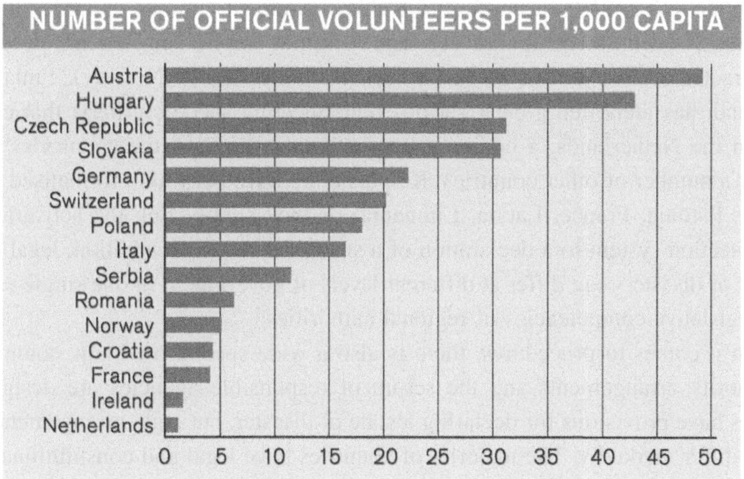

Fig. 3.3 Official volunteers per 1000 capita in Europe. *Source* Boin et al. (2014), p. 8

countries, in all cases civil security governance also comprises other actors than just the state and its institutions…especially in exceptional crisis situations' (Fig. 3.3).[23]

The involvement of the private sector in civil security matters is somewhat less prevalent in most countries, which hesitate to outsource core crisis management tasks to profit-based entities. The role of private companies in civil security matters mainly lies in the provision of local safety and emergency management, e.g. critical infrastructure protection, cybersecurity, and logistics.[24]

Concerning the role of the military in domestic disaster management, the situation is quite diverse.[25] Military forces regularly contribute to civil security efforts in support of civilian authorities in all 22 countries, at least in exceptional and prolonged crises. However, in countries lacking a strong military tradition such as the Netherlands, military assistance during domestic disasters is rare. Other countries, however, show a very strong civil-military cooperation and frequently resort to the military in domestic disasters (Italy, Norway, Sweden, Slovakia). For historical reasons, some countries established high thresholds for military involvement in civilian crisis management (Germany). Some countries have dedicated military units to support civilian authorities in case of major emergencies (Spain).

While European countries share a basic understanding of what constitutes a crisis, the points actually triggering an intervention of their civil protection authorities are very

23 Brazova (2021: 52), see also Laakso (2019), Boin et al. (2014).

24 Boin et al. (2014, p. 8), Sigala and Wakolbinger (2019).

25 Kuipers et al. (2015, p. 9).

different across countries, according to Kuipers' findings.[26] Some countries like Croatia, Ireland, or Sweden use comparatively vague criteria, emphasising the unforeseen or sudden character of an event or an 'uncontrolled development' (Norway). Finland, on the other hand, has identified a clear set of events meeting specific criteria that constitute a crisis. In the Netherlands, a crisis must constitute a threat to the country's 'vital interest'. For a number of other countries, Kuipers et al. have not found formalised definitions (Austria, Estonia, France, Latvia, Lithuania) or they simply link the activation of their civil protection system to a declaration of a state of disaster. In addition, legal definitions of crises or disasters can differ at different levels of government in one single country due to the legislative competencies of regional authorities.

When it comes to procedures, there is also a wide spectrum of how countries' internal authority arrangements and the set-up of responsible agencies are designed. Most countries have provisions for declaring a state of disaster, but legal arrangements and procedures differ markedly. The majority of countries have legal and constitutional solutions in place.

Authority arrangements for operational crisis management are typically characterised by a shared responsibility of several local agencies and emergency responders from different professional backgrounds (medical, firefighters, police), although lead agency and collaboration arrangements can differ per country. Most countries have a lead agency for overall coordination, while others rely on ad hoc networks. 17 of the studied countries created single lead agencies with integral responsibility for crisis management at national level, but with varying executive responsibilities and competences: In Croatia and Serbia, the National Rescue Board/Sector for Emergency Management are in charge of overall coordination at central level. In Hungary, the National Directorate-General for Disaster Management has a 'supervisory competence' for other agencies. In Norway and Sweden, there are national lead agencies, but their power tends to be more constrained. In Germany, the lead for natural disasters is at the level of the regional States (Länder), while the lead for civil defence disasters (armed conflict) is at the federal (national) level.[27]

As regards ensuring quality and performance within national systems, some countries have developed strong professional protocols, including frequent evaluations and inquiries (Netherlands, Sweden). However, in most countries, the number of inquiries is relatively low due to a small number of crises (Malta), legalistic, consensus-oriented political and administrative traditions (Switzerland) or a delicate balance between political actors (Serbia). Some countries have established sophisticated quality management tools (Norway, Sweden, Poland, Germany).

Interestingly, measuring the cost efficiency of national security systems seems to be one of the most under-developed aspects in national civil security systems. Governments

[26] Ibid., p. 10.
[27] Ibid., p. 12.

seem to have difficulties to get a clear overview of the spending on civil security and crisis management due to the cross-cutting nature of tasks and multi-level financing schemes.[28]

In general, national civil protection authorities enjoy broad support and legitimacy from the political and judicial system, but also from the population. Civil protection authorities receive the highest popular support in Finland, Sweden, Austria, and the Netherlands. Citizens in the 'new' Member States are, on average, more sceptical about whether their governments are doing enough to prepare for and manage disasters.

In conclusion, the ANVIL-project identified marked differences in administrative responsibilities, legal frameworks, and operational practices due to the specific national contexts and history. However, the project results did not imply drastic differences in effectiveness, efficiency, or legitimacy. Despite certain shortcomings in specific domains, the project could not identify a single best or 'one-size-fits-all' model for civil security in Europe. The authors do not consider this diversity or hybrid systems an inherent obstacle to efficient, high-quality crisis management. Citizens report a sense of security and overall trust that their systems can handle security challenges reasonably well.[29]

Identified areas for improvement include, in particular, the need to enhance cooperation across the administrative layers of government planning and preparation, forecasting and warning, control and oversight, strengthening expertise at local level, financial transparency, and enhancing public awareness.[30] At EU level, Kuipers et al. expect that increasing regulation and harmonisation of safety standards might lead to further operational harmonisation of crisis management across Member States, but the EU should only initiate activities that Member States cannot or will not do alone.[31] Boin et al. remind decision-makers that EU policies and intervention options have to connect to the existing arrangements at national level ('learning from below'), particularly of current systems enjoying strong support among the population.[32]

When looking at civil protection at EU level in the following section, this diversity should be kept in mind as it explains some of the particularities of the Mechanism. Although the introduction of the qualified majority voting system has strengthened the role of the Commission, Member States play a pivotal role in decision-making on legislation,[33] policy development, and budget at European level.

---

[28] Kuipers et al. (2015, p. 14).

[29] Boin et al. (2014, p. 3), Parker et al. (2019).

[30] Esposito (2017).

[31] Kuipers et al. (2015, p. 17).

[32] Boin et al. (2014, p. 14).

[33] The rights and obligations of the Participating States in the Mechanism are more limited. For an annual fee, they can participate in all activities of the Mechanism, but not in its formal decision-making procedures.

3.4 Disaster (Risk) Management at European Level

This section will dig into the structures, procedures, and tools of the Mechanism and also provide some illustrative examples of how its operational practice has evolved over time. Although the Mechanism is probably one of the most effective instruments for disaster management at European scale, there are also other instruments enhancing the disaster (risk) management capacity of the EU. The most important instruments in this respect are the Flood Directive, the EU Forestry Strategy, the Seveso Directive, the Nuclear Safety Directive, the Marine Strategy Framework Directive (MSFD), and the European Maritime Safety Agency (EMSA). These will be briefly presented in Sect. 3.4.5.

3.4.1 The Union Civil Protection Mechanism

Before delving into the current legislative set-up, instruments, and operational practice of the Mechanism, we will briefly recapitulate its origin, rationale, and underlying principles. Consideration of this broader historical and political context will contribute to a deeper understanding of the way in which the Mechanism operates today.

Civil protection is firmly anchored in the core of any country's national sovereignty, the duty to protect the life and well-being of its citizens. Experience shows, however, that major disasters can easily overwhelm national response capacities and transcend national borders. Large-scale disasters, therefore, provide strong incentives for international cooperation.

Countries overwhelmed by disasters often turn to their neighbours for help or to coordinate their respective response capacities as part of their risk management strategies. Most countries in Europe have bilateral agreements in that respect. Such agreements can, to some extent, provide solutions for comparatively small-scale ad hoc cross-border assistance. However, bilateral approaches quickly reach their limits when it comes to efficiently mobilising large numbers of highly specialised capacities within very short time-windows or when disasters simultaneously affect or threaten several neighbouring countries. Direct neighbours may not be able to assist because they require their national response capacities to manage the disaster on their own territory.

Furthermore, disaster management entails more than emergency response alone. It involves forecasting and early warning, as well as situational analysis, information sharing, and evaluation. Cooperation between EU Member States allows for the exchange of technical expertise and the saving of time and resources in the development of effective resources.

Multilateral frameworks like the Mechanism have distinct advantages because assistance can be drawn from a larger pool of countries through a central coordinating structure without having to negotiate with numerous countries individually. At the same time, multilateral coordination reduces the potential for duplication of efforts, particularly when it comes to assistance provided to third countries. It is therefore not surprising that disaster management became a topic on the agenda of European integration.

3.4.2 Origin of the Mechanism and Evolution of Its Legal Framework

The origin of the Mechanism goes back to the 1980s.[34] Europe had been battered by major environmental disasters with serious cross-border effects, such as the 1986 Chernobyl accident or major oil spills in coastal areas (e.g. Amoco Cadiz tanker accident in 1978). The conclusion drawn by Member States from these events was that they needed closer cooperation at the European level on civil protection matters. At the initiative of Italy, Rome hosted an informal Ministerial meeting on 2–3 May 1985, where Member States agreed for the first time to coordinate their national civil protection capacities in case of major natural disasters.[35] Ministers tasked the Commission to examine the possibility of cooperation between the Member States. They established a high-level working party as a channel for regular information exchange.

Further steps followed in March 1986 when an inventory of existing bilateral and multilateral cooperation was launched. In April 1987, the Commission presented detailed proposals for further cooperation, notably a 'Guide to Civil Defence' containing information on existing agreements, warning systems, relief plans and communication systems, contact addresses, and other relevant topics.[36] Between 1985 and 1994, six Ministerial Resolutions were adopted.[37] They resulted in the creation of operational instruments covering both the preparedness of those involved in civil protection and the response in the event of a disaster. These instruments included training, exercises, workshops, operational manuals, information exchange on 24/7 contact points, secondment of experts, etc. Initial activities focused on managing large-scale natural disasters in Europe to protect its environmental, human, and commercial interests. At Commission level the portfolio was attributed to the Directorate-General for Environment.

The attempts to establish civil protection cooperation as a policy field at European level were initially hampered by the lack of a clear legal base in the Treaties. This gradually changed with the Maastricht Treaty in 1992.[38] For the first time, civil protection was mentioned as an area in which the Community was mandated to carry out activities, although it remained unclear what kind of measures the EU could take and which

[34] Villani (2017) provides a good overview on the origin of the Mechanism.

[35] At the time the term 'civil defence' was more frequently used than 'civil protection'. Cf.: Alexander (2002), Villani (2017, p. 126).

[36] Commission of the European Communities (1987).

[37] European Commission, Directorate-General for Environment (2000).

[38] Treaty on European Union (1992).

objectives should be pursued. Activities were based on the flexibility clause of the Treaty (Article 308).

From 1998 onwards, cooperation on civil protection was gradually more formalised: the Council had adopted a Decision on 19 December 1997 establishing a Community action programme in the domain of civil protection.[39] The programme was renewed in 1999.[40] Encouraged by the success of this programme and in light of further devastating disasters hitting the continent, the Commission in September 2000 proposed the adoption of a decision establishing a community mechanism for the coordination of civil protection interventions in the event of emergencies. Accelerated by the 9/11 events in New York,[41] the Council adopted Decision 2001/792, launching the first formal legislation to facilitate reinforced cooperation between the Community and the Member States in the area of civil protection.[42]

The new Mechanism aimed at supporting and complementing Member States' activities at the national and sub-national levels through cooperation projects focusing on preparedness activities, dissemination of information, and public awareness regarding both natural and technological disasters. It also established the Monitoring and Information Centre (MIC) as the nucleus of a 24/7 coordination structure between Member States and the Commission within the Mechanism.

3.4.2.1 Council Decision 2001/792 EC

Although an important milestone in the history of the Mechanism, Decision 2001/792/EC was fraught with significant limitations. These were partly inherent in the legal instrument itself, partly because the Mechanism was largely unknown, and appropriate coordination mechanisms with other international actors such as the UN did not exist yet. Effective operations were also hampered by the fact that the coordination of Mechanism interventions outside the EU was entrusted to the Member State holding the Council Presidency of the EU (Article 6). This required time-consuming discussions between the Presidency and the Commission. Lack of means for transport was another important shortcoming: while Member States were able to offer ad hoc assistance from their warehouses or stocks, many lacked the financial and logistical means to actually transport the assets to distant disaster areas. A third important shortcoming was the lack of early warning systems. This became

[39] Council Decision of 19 December 1997 establishing a Community action programme in the field of civil protection. 98/22/EC. OJ L 8 (14 January 1998). https://eur-lex.europa.eu/legal-content/EN/TXT/PDF/?uri=CELEX:31998D0022&qid=1741340179524.

[40] Council Decision 1999/847/EC of 9 December 1999 establishing a Community action programme in the field of civil protection. OJ L 327, (21 December 1999). https://eur-lex.europa.eu/legal-content/EN/TXT/PDF/?uri=CELEX:31999D0847.

[41] Morsut (2014, p. 145).

[42] Council Decision of 23 October 2001 establishing a Community mechanism to facilitate reinforced cooperation in civil protection assistance interventions; 2001/792/EC, Euratom, OJ L 297/7 (15 November 2001). https://eur-lex.europa.eu/legal-content/EN/TXT/PDF/?uri=CELEX:32001D0792.

acute in the wake of the 2004 Indian Ocean tsunami when the Mechanism, like the rest of the international disaster management community, was caught largely unprepared for a disaster of such magnitude. The lessons from these and other cases ultimately led to a review of the legislation in 2007.

3.4.2.2 Council Decision 2007/779/EC, Euratom

Council Decision 2007/779 contained improvements in many respects.[43] Six key measures can be highlighted: 1) the development of pre-defined modules to form a European rapid response capability,[44] 2) contributing to the further development of early warning systems, 3) improving the communication within the Mechanism through the development of a common emergency communication and information system, 4) enhancing the availability of transport resources through pooling and co-financing, 5) reinforcing the mandate of the Mechanism to include support for consular assistance; and 6) transferring operational coordination of Mechanism deployments to the Commission's MIC. The Presidency maintained an overall coordination role at the political and strategic levels.

The MIC's growing operational experience and a rigorous 'lessons learnt' programme introduced in 2006 led to further improvements. The difficult relationship between the Mechanism as a new player on the international disaster management scene and the established UN system was increasingly improved through confidence-building measures such as joint assessment missions of UN and EU Civil Protection Teams, e.g. during the floods in Namibia in 2009.

In parallel to this, internal coordination between the Commission's humanitarian service (ECHO) and its civil protection branch in DG Environment was strengthened. Both entities were merged under the umbrella of one Directorate-General (DG ECHO) and one Commissioner in 2010. Partnering two systems with similar aims, but different operational philosophies posed some challenges, but ultimately yielded important complementarities and synergies.[45] For instance, the monitoring and information tools of the MIC could also be used by the humanitarian branch, and civil protection benefitted from the expertise of the humanitarian field network.

Experience from several challenging deployments of EU Civil Protection Teams (EUCPT) inside and outside Europe led to further changes in the *modus operandi* for

---

[43] Council Decision of 8 November 2007 establishing a Community Civil Protection Mechanism ('recast'); 2007/779/EC, Euratom, OJ L/314/9 (01 December 2007). https://eur-lex.europa.eu/legal-content/EN/TXT/?uri=celex%3A32007D0779%2801%29.

[44] Article 6 of Decision 1313/2013/EU of the European Parliament and of the Council of 17 December 2013 on a Union Civil Protection Mechanism, OJ L 347 (17 December 2013) defines a module as '...a self-sufficient and autonomous predefined task- and needs-driven arrangement of Member States' capabilities or a mobile operational team of the Member States, representing a combination of human and material means that can be described in terms of its capacity for intervention or by the task(s) it is able to undertake' https://eur-lex.europa.eu/eli/dec/2013/1313/oj/eng.

[45] Some humanitarian actors, for instance, were concerned that civil protection might 'politicise' assistance and jeopardise adherence to the humanitarian principles.

such missions, in particular in relation to reinforcing preparedness for security incidents and media relations. Media relations had been particularly challenging, for instance, during the EUCPT mission deployed to Taiwan after Typhoon Morakot in 2009. As a lesson from these deployments, several measures were introduced to better prepare EUCPT for potentially sensitive missions, such as pre-established press briefings and media training.

During a deployment in support of the International Organization for Migration (IOM) to a refugee camp in the border area between Libya and Tunisia in 2011 the EUCPT experienced a serious security incident. Rioting third country nationals attacked and looted the IOM office at the very moment when one of the EUCPT members had a meeting with IOM staff. This incident triggered reinforced security management protocols for missions.

The EUCPT mission to Japan after the tsunami of 11 March 2011 also proved highly complex, forcing the MIC to resort to improvised solutions for local transport and logistics, for business continuity due to the time difference between Europe and Japan, and to ensure permanent availability of nuclear expertise.

A key lesson from the deployment to Japan was that, although the mission reached its objectives, too many issues had to be improvised. As a direct consequence, three main improvements were proposed for discussion in the legislative revision in 2013: 1) to include a financing facility covering local transport and logistics in the legislation; 2) to enable the MIC to sustainably cope with demanding emergency situations in different time zones through appropriate equipment and permanent staffing; and 3) to ensure availability of technical expertise regarding specific hazards and risks at any moment. How these lessons were implemented will be shown in the following sub-section.

3.4.2.3 Decision 1313/2013

The legal revision proposed by the Commission and finally adopted with some modifications as Decision 1313/2013 contains several improvements compared to the 'recast' of 2007.[46]

First, the MIC was transformed into the Emergency Response Coordination Centre (ERCC). Apart from moving to new premises, the main change consisted in the establishment of a full 24/7 duty system, with a minimum of 2 duty officers present in the Centre around the clock.

Second, the availability of rapidly deployable capabilities was enhanced by the creation of the European Emergency Response Capacity (EERC). The EERC consisted of a voluntary pool of pre-committed assets which Member States committed to make available in case of a request addressed to the Mechanism. To ensure high quality of the capacities committed to the pool, a certification system and a co-financing scheme to adapt capacities to comply with the required technical standards were introduced. The Commission

[46] Decision 1313/2013/EU of the European Parliament and of the Council of 17 December 2013 on a Union Civil Protection Mechanism, OJ L 347 (17 December 2013). https://eur-lex.europa.eu/eli/dec/2013/1313/oj/eng.

was also mandated to establish capacity goals for the voluntary pool. The EERC was renamed European Civil Protection Pool (ECPP) in 2019.

The *third* important amendment concerned transport and logistics. The transport co-financing rate was increased from 50% to 55% for ad hoc offers and to 85% for capacities committed to the voluntary pool as an additional incentive for Member States to register assets in the pool. Furthermore, costs for local transport, short-term storage, and repackaging were 100% covered by the Union budget.

A *fourth* important change concerns the possibility given to some international organisations to request assistance via the Mechanism (UN, IFRC, Organization for the Prohibition of Chemical Weapons, International Atomic Energy Agency). This measure was introduced to further promote consistency in the response to disasters outside the EU. With Article 26 the 2013 revision also included an explicit commitment of UCPM operations to the humanitarian principles.

The revision also marked an important change concerning its legal base in the European Treaties: In the 2001 and 2007 versions of the Mechanism, reference was still made to the 'Flexibility Article' (Article 308) in the Maastricht Treaty. In the meantime, civil protection (and humanitarian aid) had been explicitly established as Community policies in the Lisbon Treaty.[47] Articles 6f and 196 of the Lisbon Treaty are the legal base for civil protection and Article 214 for humanitarian aid. Article 222—the so-called 'Solidarity Clause'—introduced an obligation for Member States to mutually assist each other in case of disasters (see Sommario's chapter in this volume). The Lisbon Treaty also introduced qualified majority voting into the civil protection legislation, which strengthened the role of the Commission and conferred a stronger mandate to the Mechanism as decisions could no longer be blocked by individual countries.

In parallel with the enhanced competences, the budget increased: the first Civil Protection Action Programme in 1998 had a financial envelope of 1.5 million European Currency Unit (ECU, equivalent to 1.5 million EUR) per year. The financial envelope of the 2001–2007 legislation grew to 11 million EUR, then climbed to 189 million EUR for the period 2007–2013. From 2014 to 2020, the Mechanism budget reached 368 million EUR, and for 2021 to 2027 further increased to 3.7 billion EUR.[48]

Decision 1313/2013 is still the current legal basis of the Mechanism, although there have been significant amendments. The year 2019, for instance, marked the initiation of the Union Civil Protection Knowledge Network (UCPKN), moulded into an Implementing Act 2 years later.[49] Apart from grouping existing programmes like training, exercises, and the exchange of experts under a coherent umbrella, a key aim of the UCPKN was

[47] Treaty on the Functioning of the European Union (2012).

[48] European Commission (2024b, p. 15).

[49] Decision (EU) 2019/420 of the European Parliament and of the Council of 13 March 2019 amending Decision No 1313/2013/EU on a Union Civil Protection Mechanism, OJ L77 I/1. https://eur-lex.europa.eu/eli/dec/2019/420/oj/eng; Commission Implementing Decision (EU) 2021/1956 of 10

to strengthen the Mechanism's scientific pillar, i.e. to engage deeper with the scientific community and make more practical use of their findings and methods.

The main modification adopted in 2019 concerned the creation of a reserve capacity called rescEU. rescEU complements the European Civil Protection Pool (ECPP) with a layer of additional response capacities entirely financed from the Community budget.

The rescEU initiative was triggered by the devastating fire season of 2017, which had caused 117 deaths in Portugal. As many countries were affected by strong fires almost simultaneously during that summer, there were not enough aerial firefighting capacities available to respond to all requests for assistance. Out of a total of 17 requests during that summer, assistance could actually be delivered in only 10 cases.[50] The 2017 forest fire season triggered a legislative proposal from the Commission aiming at the creation of a reserve capacity for aerial firefighting capabilities. The proposal was adopted by Member States and the European Parliament with minor modifications in 2019. The rescEU concept proved successful and was extended subsequently during the COVID-19 pandemic and the war in Ukraine.

The reserve capacities were established through Implementing Decisions of the Commission.[51] During COVID-19, the Commission adopted several such Decisions between 2019 and 2022 to create medical reserves and medical evacuation capacities under rescEU. The concept was further elaborated following the Russian war of aggression against Ukraine, which triggered the establishment of reserve capacities for shelter, emergency energy supply, CBRN (Chemical, Biological, Radiological and Nuclear) detection, and logistics. Within the context of rescEU, Member States also agreed to mandate the Commission to directly procure assistance in specific cases of urgency.

3.4.3 Current Legal Framework of the Mechanism

For the time being, Decision 1313/2013 remains the core legislative basis of the Mechanism. The Decision contains seven chapters covering the objectives and scope of the legislation, activities in prevention, preparedness, and response, as well as financial, general, and final provisions. The general objective set out in Article 1 of Decision 1313/2013 is to strengthen cooperation between the Union and Member States to improve the effectiveness of systems for preventing, preparing for, and responding to natural and

November 2021 on the establishment and organisation of the Union Civil Protection Knowledge Network, OJ L 399/1 (10 November 2021). https://eur-lex.europa.eu/legal-content/EN/TXT/PDF/?uri=CELEX:32021D1956.

[50] European Commission (2017, p. 3).

[51] Most of these Acts were recently grouped together in one new document: Commission Implementing Decision (EU) 2025/704 of 10 April 2025 laying down rules for the implementation of Decision No 1313/2013/EU of the European Parliament and of the Council on a Union Civil Protection Mechanism and repealing Commission Implementing Decisions 2014/762/EU and (EU) 2019/1310; https://eur-lex.europa.eu/legal-content/en/TXT/?uri=CELEX%3A32025D0704.

man-made disasters. The Decision aims, in particular, to protect people against various disasters, but its scope also covers the environment, property, and cultural heritage. It fosters high-level protection against disasters, promotes public awareness and preparedness, cooperation at the cross-border level, and efficient disaster response. The geographical scope of the Mechanism is global, it covers disasters both inside and outside the EU.

Article 28 of Decision 1313/2013 stipulates that the Mechanism shall be open to the participation of countries belonging to the European Free Trade Association (EFTA) and to acceding countries, candidate countries and potential candidates for EU Membership. Ten such countries currently participate in the Mechanism in addition to the 27 Member States. The legislation pursues a comprehensive concept of disaster management, covering prevention, preparedness, and response aspects in an integral manner.

International arrangements are usually characterised by a mix of rights and obligations for the contracting parties. This equally applies to the UCPM. Decision 1313/2013 assigns different roles and responsibilities in each phase of a disaster to both key actors in the system, the European Commission and Member States.

3.4.3.1 Prevention

The Commission's responsibilities in relation to prevention are enshrined in Article 5 of Decision 1313/2013. Essentially, its role can be characterised as that of a facilitator. It makes use of a broad catalogue of instruments and actions aiming to improve the overall knowledge base on disaster risks, facilitating and promoting cooperation and knowledge-sharing on results of scientific research and innovation, best practices, information dissemination, reporting, and exchange of experience. Activities cover relevant domains such as risk management, risk assessment and mapping activities, cross-sectoral overviews, and maps of natural and man-made disaster risks. The Commission also has a role in promoting the use of funding opportunities for prevention measures available at EU level, including its calls for proposals such as the Knowledge for Action in Prevention and Preparedness (KAPP).

At the request of a Member State, a third country, the UN or its agencies, the Commission may also deploy expert teams to provide advice on prevention measures. Seven such 'advisory missions' have been deployed to different countries inside and outside the EU in the period 2022 to 2024.[52]

The key Instruments of Member States in the *prevention* phase (Article 6) mainly encompass the development of national and sub-national risk assessments, as well as developing and refining their disaster risk management planning. On a voluntary basis, they can also have their risk management capabilities peer reviewed by other Member States.

[52] ECHO Daily Map of 08/08/2024. https://erccportal.jrc.ec.europa.eu/ECHO-Products/Maps#/maps/4961.

3.4.3.2 Preparedness

While the Commission mandate in the prevention area includes mainly 'soft instruments', often without binding character, its role in the preparedness phase (Articles 7–13 of Decision 1313/2013) is much more robust. Its key responsibility is to establish and run the ERCC. The ERCC is the Mechanism's central coordination hub. It is the key instrument of the Mechanism to coordinate, monitor, and support in real time the response to emergencies at Union level and operates in close contact with national civil protection authorities and relevant Union entities promoting cross-sectoral cooperation.

In addition, the Commission is held to implement general preparedness actions, notably to run the Common Emergency Communication and Information System (CECIS), the central communication backbone linking the ERCC with the national contact points in Member States. It also ensures comprehensive access to science-based alerts, monitoring and analytical tools through networking and partnering up with research institutions. In terms of operational preparedness, the Commission defines the quality requirements and organises the certification for the capacities pre-committed by the Member States to the ECPP according to Article 11.

Article 12 confers on the Commission the mandate to define the rescEU capacities and the capacities needed ('capacity goals'), based on existing scenarios, identified and emerging risks, overall capacities, and gaps at EU level. rescEU capacities currently cover aerial forest-firefighting, chemical, biological, radiological and nuclear incidents, emergency medical response, shelter, transport, and logistics. In practical terms, the Commission concludes operational contracts with the Member States that own, rent, or lease such capacities. These contracts specify the terms and conditions of deployment of rescEU capacities. The rescEU capacities are available for response operations under the UCPM following a request for assistance through the ERCC. The decision on their deployment and demobilisation, including in the event of conflicting requests, is taken by the Commission in close coordination with the requesting Member State and the Member State owning, renting, or leasing the capacity. The acquisition, renting, or leasing of the rescEU capacities is financed through direct grants from the EU budget.

The Commission also establishes and manages the mobilisation and dispatch of expert teams from a network of trained experts to support assessment and coordination of the disaster management efforts on site or to provide logistical support (Article 8). It also has general tasks in terms of ensuring a high level of operational preparedness, including developing host nation support guidelines, facilitating pre-positioning of response assets, improving the interoperability of modules, and deploying expert teams to provide advice on preparedness measures.

Last, but not least, the Commission is responsible for the UCPM Knowledge Network (Article 13). The network aims to aggregate, process, and disseminate knowledge and information relevant to the UCPM in a multi-hazard approach including relevant civil protection and disaster management actors, centres of excellence, universities, and researchers.

The key requirement for Member States in the preparedness phase (Article 9) is to designate permanent national contact points (NCPs). NCPs are the indispensable link to facilitate communication and coordination between the national civil protection systems and the ERCC. In addition, Member States ensure appropriate preparedness actions to facilitate host nation support. Some preparedness actions of Member States are, in principle, obligatory, but can be implemented on a voluntary basis. This concerns, in particular, the establishment of response capacities such as modules, other capacities, and experts committed to the EU reserve capacity or the ECPP (Article 12). Once a capacity is committed to the Pool, deployment upon request of the UCPM becomes, in principle, obligatory although the Member State maintains the final decision and can opt out in cases of *force majeure*. For rescEU capacities, however, the opt-out possibility does not exist.

3.4.3.3 Response

During the *response* phase (Articles 14–18 of Decision 1313/2013), a Member State affected by a disaster or imminent disaster immediately notifies other Member States and the Commission (ERCC), in particular if the disaster is causing multi-country transboundary effects or affects other Member States. The Member States to which a request for assistance is addressed via the UCPM then decide whether they are in a position to offer assistance.

In the case where assistance is provided, the requesting Member State is responsible for providing host nation support and directs the incoming assistance. Member States ensure appropriate information flows about their assistance. They play a pivotal role in any UCPM operation since practically all the capacities are provided by them, starting from the ad hoc donations of assistance (e.g. shelter, medicines), to the deployment of specialised teams (e.g. Urban Search and Rescue, USAR), pre-committed capacities from the ECPP or rescEU, or members of the EUCPT deployed to the disaster theatre.

The main role of the Commission (Article 15) in this process is to facilitate coordination, to ensure consistency and overall effectiveness of the Mechanism operations. For this, the ERCC has several instruments at its disposal. As soon as the Mechanism is activated through a request for assistance, the ERCC liaises with the contact points of other Member States. In cooperation with the affected country the ERCC establishes an analysis of the situation and the response needed, based on the identified needs and relevant pre-developed response plans. The ERCC may also invite Member States to deploy rescEU assets in cases where ad hoc assistance and ECPP capacities are not sufficient to effectively address a disaster.

When a disaster occurs outside the EU, the affected country or one of the international organisations mentioned in Sect. 3.4.2.3 can direct a request for assistance to the ERCC. The ERCC immediately disseminates the request to the Member States. They decide on the assistance they are prepared to offer (Article 16). The ERCC also deploys an EUCPT on site to facilitate coordination of incoming EU assistance on site and to support the relief effort through situational assessment and/or specific technical expertise. The EUCPT is composed of experts from Member States and one or two liaison officers from the ERCC. Mechanism operations under Article 16 may be conducted either as an autonomous assistance intervention or as a contribution to an intervention led by the UN or another international organisation. In such cases, the Mechanism respects the lead role of the United Nations Office for the Coordination of Humanitarian Affairs (OCHA) and fully integrates into its overall coordination system. In case a Mechanism intervention is conducted in the context of a man-made disaster or complex emergency, consistency with the European Consensus on Humanitarian Aid and respect for humanitarian principles is ensured. The basic modus operandi of a Mechanism response operation is described in Fig. 3.4.

The ERCC can also support coordination through its transport instruments (Article 18). It provides and shares information on equipment and on transport/logistical resources available in Member States (e.g. pooling of transport capacities through joint use of an aircraft). It can also assist Member States through identifying transport and logistical resources or equipment that may be available from other sources (e.g. through a framework contract with a transport broker).

In terms of financial support, there is a progressive involvement and funding by the EU: Member States offering assistance on an ad hoc basis can request the Commission to

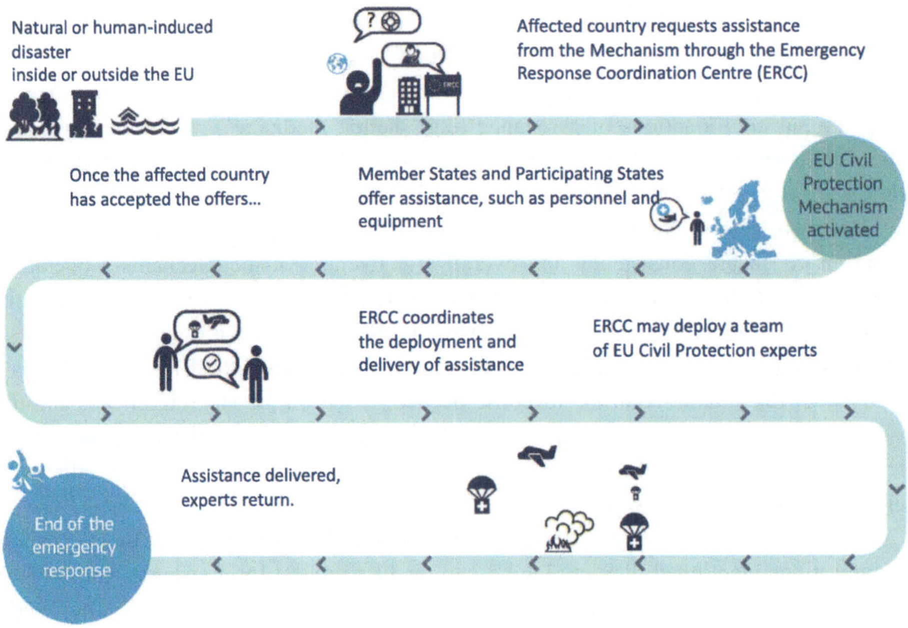

Fig. 3.4 Modus operandi of the UCPM. *Source* DG ECHO

reimburse 75% of the international transport costs and 100% of the local transport costs. For capacities deployed from the ECPP, the Commission can not only cover 75% of the transport costs, but also 75% of the operational costs. In case rescEU assets are deployed, up to 100% of the transport and operational costs can be covered by the Commission. The amounts provided by the ERCC to support Member States' transport activities reached ca. 90 million EUR in 2022.

3.4.4 Operational Experience of the UCPM—Current Practice and Trends

Over the almost 25 years of its existence, the UCPM was able to gain ample operational experience. At the time of writing, the UCPM has been activated more than 800 times since its inception. The number of requests for assistance per year has steadily increased: from three in 2002 to more than 200 in 2022. Over a long period, the number of requests oscillated around 20 per year, but the number exponentially increased during the COVID-19 pandemic and the Ukraine war. With the decline of COVID-19 and more compact requests coming from Ukraine, the overall number of requests decreased, but remained significantly above the figures for the period between 2005 and 2019 (Table 3.2).

Comparing activations of the Mechanism inside the EU and outside shows that the distribution is more or less balanced (Table 3.3). To be noted that the European Commission currently does not maintain a single centralised, methodologically uniform and publicly available database allowing to cleanly split the 800+ UCPM activations by individual year for inside/outside EU or disaster categories throughout the complete period. Figures may thus have to be slightly revised once consolidated data become available

Looking at the disaster types for which the Mechanism has been activated over the period 2007 to 2024, the majority of requests concern pandemic outbreaks, in particular related to COVID-19. The more 'classical' natural disasters such as wildfires, floods, storms, and earthquakes also rank among the top 10, as well as consular support and support for human disasters like population displacements, as can be seen in Table 3.4.

Table 3.4 provides a static picture of the disaster landscape. In recent years, however, that landscape has considerably evolved. From 2017 to 2022, there has been a significant shift to pandemic response and man-made disasters. Requests for assistance related to medical assistance during the COVID-19 pandemic accounted for 57% of all requests in 2020 and even 61% in 2021 before decreasing to 19% the year after. In 2022, the war in Ukraine accounted for 54% of all activations.[53]

Environmental and natural disasters thus remain important activities, but there has been a gradual extension of Mechanism interventions towards other, more complex types of disasters (health emergencies, armed conflicts). These activations pose more challenges to the UCPM, for instance, in relation to increasingly difficult security contexts and complexity

[53] European Commission (2024b, p. 13 ff).

Table 3.2 Requests for assistance to the UCPM 2002–2025

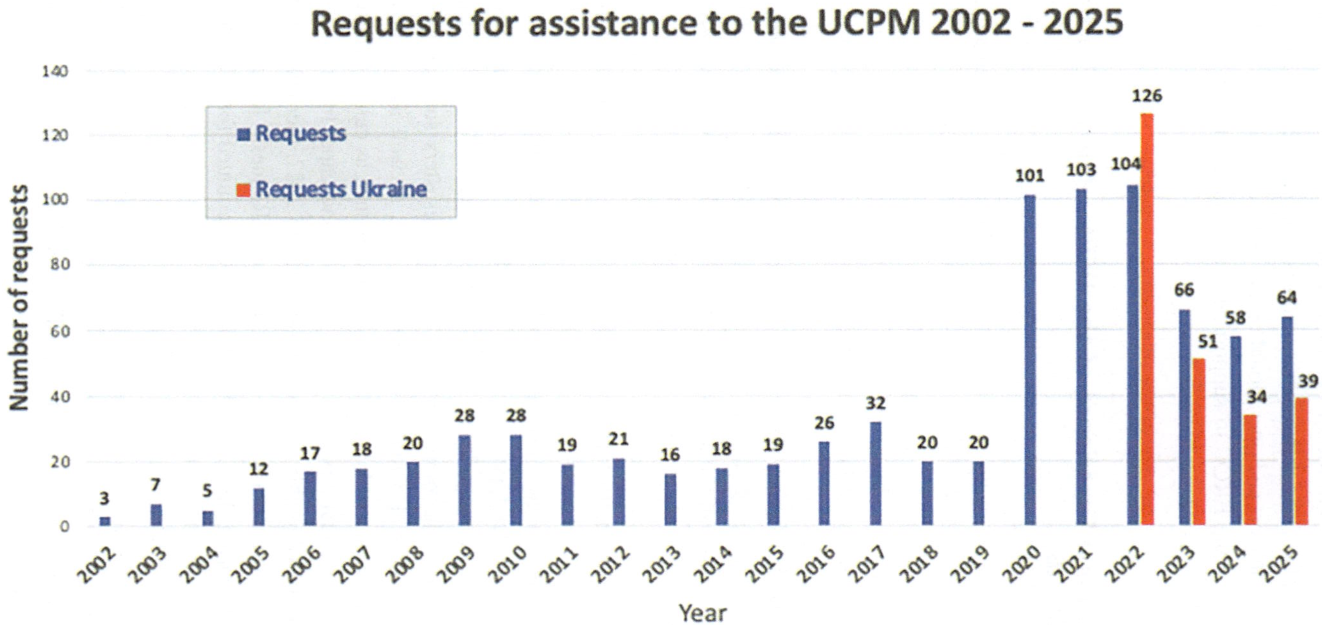

Source preliminary data kindly provided by DG ECHO and slightly re-arranged by the author. Updated requests from Ukraine displayed separately to highlight scale of emergency needs

Table 3.3 Requests for assistance inside and outside the EU 2007–2025

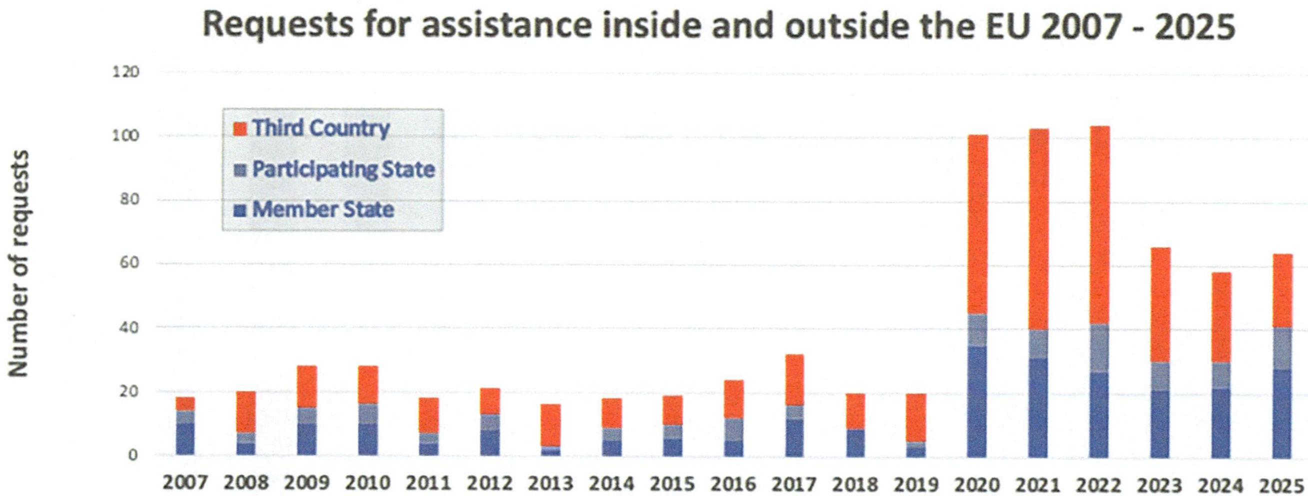

	2007	2008	2009	2010	2011	2012	2013	2014	2015	2016	2017	2018	2019	2020	2021	2022	2023	2024	2025
Third Country	4	13	13	12	11	8	13	9	9	12	16	11	15	56	63	62	36	28	23
Participating State	4	3	5	6	3	5	1	4	4	7	4	0	2	10	9	15	9	8	13
Member State	10	4	10	10	4	8	2	5	6	5	12	9	3	35	31	27	21	22	28

Year

Source preliminary data kindly provided by DG ECHO and slightly re-arranged by the author. Updated requests for assistance from Ukraine not included

Table 3.4 Requests for assistance to the UCPM by disaster type 2007–2025

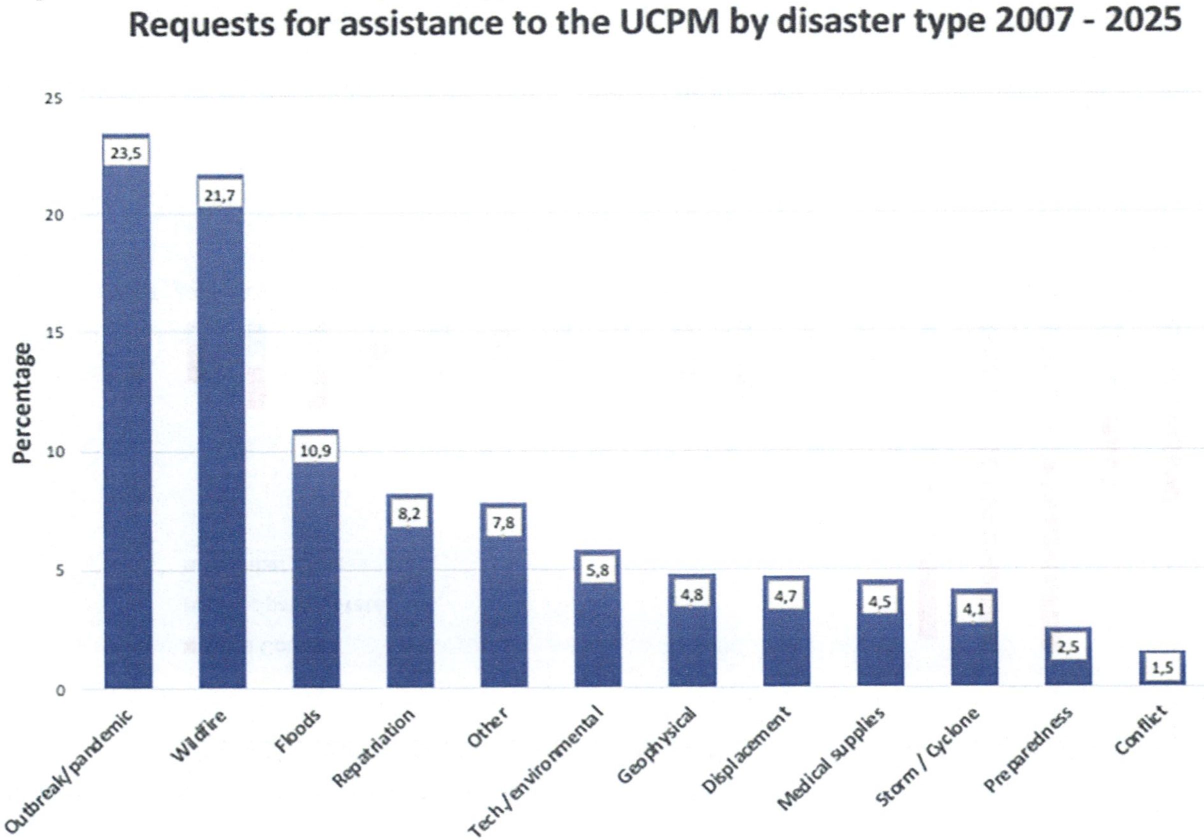

Source preliminary data kindly provided by DG ECHO and slightly re-grouped by the author. Updated requests for assistance from Ukraine not included

of operations (e.g. medical evacuations). In parallel, the number of activations has significantly increased, which requires adequate human and financial resources. Furthermore, the COVID-19 and Ukraine operations were characterised by their long durations, constituting additional challenges in terms of operational resilience and business continuity.

The Ukraine war triggered the largest and most complex UCPM operation to date. Ukraine activated the UCPM on 15 February 2022 and updated its initial request multiple times. The spectrum of requests included medical supplies, food, shelter, firefighting equipment, computers and communications equipment, cultural heritage protection equipment, CBRN countermeasures, and seeds for agriculture. The ERCC supported Poland, Romania, and Slovakia to establish logistics hubs in which incoming assistance was received and consolidated before being dispatched to Ukraine. Furthermore, Belgium hosts a rescEU stockpile for specialised private sector donations (e.g. pharmaceutical products). Responding to increasing needs for medical treatment, the European Commission also set up a standard operating procedure for the medical evacuation (MEDEVAC) of persons with difficult medical conditions from Ukraine.

By February 2026, more than 11,000 power generators were offered to Ukraine through the UCPM, more than 5000 individuals were evacuated for medical treatment in the EU, and more than 150,000 tonnes of civil protection assistance were provided by 33 countries, on top of 1.4 bn 950 EUR worth of humanitarian assistance provided to Ukraine and Moldova. In addition, several EUCPT expert missions were deployed to Ukraine between 2022 and 2024 to advise the government on streamlining the MEDEVAC procedure and on CBRN matters.[54]

3.4.5 Other Disaster Risk Management Instruments at European Level

3.4.5.1 Flood Directive

The *EU Flood Directive* aims to assess and manage flood risks across Europe to reduce their impact on human health, the environment, cultural heritage, and economic activities.[55] The Directive requires Member States to implement five key measures: Member States are obliged to 1) conduct preliminary flood risk assessments to identify areas at risk; 2) create flood hazard and risk maps; 3) establish risk management plans to outline

[54] ECHO Daily Maps 08/08/2024, 07/01/2025, 16/01/2025, 24/02/2025, 02/06/2025, 11/02/2026, 04/03/2026, 10/04/2026. https://erccportal.jrc.ec.europa.eu/ECHO-Products/Maps#/maps?pageIndex=1&mt=Daily%20Map.

[55] Directive 2007/60/EC of the European Parliament and of the Council of 23 October 2007 on the assessment and management of flood risks, OJ L 288 (06 November 2007). https://eur-lex.europa.eu/eli/dir/2007/60/oj/eng.

specific measures to prevent, protect against, and prepare for floods; 4) ensure participation of the public in the planning process; and 5) update flood risk assessments, maps, and management plans every 6 years.

The Directive has improved flood risk management and preparedness across Europe in several ways: public access to information has raised awareness about flood risks and the importance of preparedness. Furthermore, measures under the Directive help protect ecosystems and biodiversity from flood damage. The Directive also yields economic benefits by reducing flood risks and helps minimise economic losses and damage to infrastructure.

Despite notable improvements, however, the 2025 report on the implementation of the Flood Directive identified scope for further improvement. It concludes, for instance, that Member States need to further enhance their planning capacity and implement measures adapting to future climatic conditions.[56]

3.4.5.2 Seveso Directive

The '*Seveso III Directive*' aims at preventing major industrial accidents involving dangerous substances and minimising their consequences for human health and the environment.[57] Named after the Seveso disaster in Italy in 1976, the Directive sets out stringent rules for industrial establishments handling hazardous substances, including reporting requirements.[58] It applies to over 12,000 industrial installations across the EU, laying down rules to prevent major industrial accidents and minimise their harmful impacts on human health and the environment.

The measures to prevent accidents can essentially be grouped in four categories: 1) prevention and safety measures requiring establishments to implement safety management systems; 2) industrial facilities have to report to national authorities if they store or use large quantities of hazardous substances; 3) the Directive identifies hazardous substances through a detailed list and generic hazard classifications; and 4) the Directive ensures that the public is informed about industrial activities and their potential risks.

The Directive seems to be generally fulfilling its purpose although leaving scope for further improvement. Recent studies state that the three EU Seveso Directives significantly improved safety in the process industries.[59] Other studies, however, are more critical, highlighting, for instance, that process safety accidents have not decreased, neither in number nor in severity.[60]

---

[56] European Commission (2025).

[57] Directive 2012/18/EU of the European Parliament and of the Council of 4 July 2012 on the control of major-accident hazards involving dangerous substances, amending and subsequently repealing Council Directive 96/82/EC, *OJ L 197 (24 July 2012)*. https://eur-lex.europa.eu/legalcontent/EN/TXT/PDF/?uri=CELEX:32012L0018.

[58] European Commission (2021a).

[59] Prerna et al. (2017).

[60] Laurent et al. (2021).

3.4.5.3 Nuclear Safety Directive

The *Nuclear Safety Directive* establishes a framework to ensure the safety of nuclear installations within the European Union.[61] It obliges Member States to establish and maintain legislative and regulatory measures for nuclear safety at national level. In addition, regulatory authorities have to be independent and functionally separate from any organisations promoting or utilising nuclear energy. In terms of safety, the Directive contains objectives to prevent accidents, mitigate their consequences, and avoid accidental radioactive releases. Furthermore, regulatory authorities and license holders have an obligation to furnish information on operating conditions, incidents, and accidents. Nuclear installations are subjected to undergo periodic safety reviews. Last, but not least, public participation in decision-making related to the licensing of nuclear installations is encouraged by the Directive. The Directive has strengthened the powers and independence of regulatory authorities, which improved nuclear safety standards. It has also improved accident management and on-site emergency preparedness and response. The establishment of a system of EU-wide peer reviews promoted improvement and sharing of good practice. In a press release issued in October 2012, the European Commission concludes that the stress-tests undertaken after the Fukushima nuclear accident confirmed a satisfactory level of safety for nuclear reactors in the EU, but also identified scope for improvement, in particular as international standards and practices have not been applied everywhere.[62]

3.4.5.4 Marine Pollution

The EU Marine Pollution Legislation primarily builds on two key pillars: the *Marine Strategy Framework Directive (MSFD)*[63] and the *Directive on ship-source pollution.*[64] These Directives aim to protect marine environments from various forms of pollution, including oil spills. The MSFD requires EU Member States to develop marine strategies to achieve or maintain 'Good Environmental Status' of their marine waters. It includes measures to prevent pollution, protect biodiversity, and manage human activities impacting the marine environment. The Directive on Ship-source Pollution aims to prevent pollution from ships by setting regulations on discharges and introducing penalties for

---

[61] Council Directive 2009/71/ Euratom of 25 June 2009 establishing a Community framework for the nuclear safety of nuclear installations, OJ L 172, (2 July 2009) (amended by Council Directive 2014/87/Euratom of 8 July 2014) https://eur-lex.europa.eu/eli/dir/2009/71/oj/eng.

[62] European Commission (2012).

[63] Directive 2008/56/EC of the European Parliament and of the Council of 17 June 2008 establishing a framework for community action in the field of marine environmental policy, OJ L 164/19 (25 June 2008). https://eur-lex.europa.eu/legal-content/EN/TXT/PDF/?uri=CELEX:32008L0056.

[64] Directive (EU) 2024/3101 of the European Parliament and of the Council of 27 November 2024 amending Directive 2005/35/EC as regards ship-source pollution and on the introduction of administrative penalties for infringements, OJ L (16 December 2024). https://eur-lex.europa.eu/legal-content/EN/TXT/PDF/?uri=OJ:L_202403101.

violations. It also includes provisions for monitoring and controlling maritime traffic to prevent accidents and spills.[65]

A key role in the prevention and response to major oil spills is also played by the European Maritime Safety Agency (*EMSA*). EMSA is a key institution in combating marine pollution. It plays a crucial role in oil pollution response by providing response vessels and services like CleanSeaNet, a satellite-based monitoring system helping to detect oil spills in near-real-time, enabling quicker and more effective responses. According to the Agency, the number of oil spills in European waters shows a positive declining trend.[66]

3.4.5.5 EU Adaptation, Biodiversity, Forest, and Soil Strategies

As part of its broader approach to the climate and environmental crises, the EU has adopted a set of interconnected strategies that contribute to building resilience and preventing disasters. The new *EU Strategy on Adaptation to Climate Change*,[67] the *EU Biodiversity Strategy for 2030*,[68] the *EU Forest Strategy 2030*,[69] and the *EU Soil Strategy for 2030*[70] form a coherent policy framework under the European Green Deal. These strategies are highly relevant to disaster management, as they seek to reduce climate- and ecosystem-related risks, enhance adaptive capacity, and promote nature-based solutions for disaster prevention.

For instance, the *EU Strategy on Adaptation to Climate Change* calls for stronger integration between climate adaptation and disaster risk reduction policies, recognising that coordinated action across sectors and levels of governance is essential to reduce climate-related risks.

The *EU Biodiversity Strategy for 2030* sets out ambitious targets to halt and reverse biodiversity loss across the Union. It aims to protect at least 30% of the EU's land and sea areas, ensure the strict protection of all remaining primary and old-growth forests, and restore degraded ecosystems. Member States are expected to implement effective restoration measures, with particular emphasis on ecosystems that have the highest potential to capture and store carbon, and to prevent or reduce the impact of natural disasters.

The objective of the *EU Forest Strategy 2030* is to enhance the quantity and quality of forests in the EU while ensuring their protection, restoration, and resilience.[71] The strategy inter alia supports the adaptation of forests to weather extremes and climate change.

---

[65] For a more detailed account, including recent measures to further strengthen the environmental protection measures of the Directive, see European Parliament Briefing (2024) EU Legislation in Progress: Revision of the Directive on Ship-source Pollution. https://www.europarl.europa.eu/RegData/etudes/BRIE/2023/753928/EPRS_BRI%282023%29753928_EN.pdf.

[66] European Maritime Safety Agency—EMSA (2025). https://www.emsa.europa.eu/protecting-the-marine-environment/oil-pullution-response.html.

[67] European Commission (2021b).

[68] European Commission (2020).

[69] European Commission (2021c).

[70] European Commission (2021d).

[71] European Commission (2021c).

By promoting sustainable forest management and reducing the accumulation of fuels, the strategy also aims to diminish the risk of wildfires. Restoring forests and increasing biodiversity can also reduce the susceptibility of forests to fires. Encouraging public involvement in forest management can lead to better fire prevention practices. As recent studies correctly conclude, however, it is too early to make any meaningful assessment.[72] Whether the strategy will actually yield the expected results remains to be seen at the end of the implementation period.

Finally, the *EU Soil Strategy for 2030* complements the forest, biodiversity, and adaptation policies by framing soils as a foundational ecosystem service critical for climate resilience and disaster risk reduction. The strategy emphasises the crucial role of soils in carbon sequestration, water regulation, and erosion control, underscoring their importance in mitigating the impacts of floods, landslides, and droughts. In July 2023, the European Commission proposed a Soil Monitoring and Resilience Directive, originally announced as the 'Soil Health Law'.[73]

3.5 Conclusion and Outlook

The preceding sections confirm that the UCPM has proven to be a versatile instrument, which has acquired ample experience and implemented numerous lessons from around 900 operations since 2001. It has significantly enhanced coordination with stakeholders, including the UN and humanitarian aid actors. With the ECPP, rescEU, and the Knowledge Network, the Mechanism has established a robust set of rapid response capacities, a large reservoir of well-trained experts and a solid monitoring and analytical base. The iterative adaptation of its legal and financial framework contributed to achieving faster, more coherent, and efficient operations, contributing to enhancing operational preparedness. The Commission's 2024 evaluation of the UCPM confirms that civil protection remains a key pillar in the EU's prevention, preparedness, and response activities to face the more frequent, intense, and complex disasters of the future.[74]

The evaluation also identified scope for improvement, mentioning, for instance, a lack of human and financial resources, overly complex administrative procedures, and too strong a focus on response. External factors like the changing threat and risk landscape, as well as gaps in the scientific evidence base, were mentioned as further limiting factors.

Despite those challenges the evaluation concludes on a positive note, stating that 1) the UCPM has shown relevance for the prevention, preparedness, and response to natural and man-made disasters and its ability to adapt to unforeseen events and cross-sectoral

[72] See, for instance, Gordeeva et al. (2022, p. 12).

[73] Proposal for a Directive of the European Parliament and of the Council on Soil Monitoring and Resilience (Soil Monitoring Law), COM/2023/416 final (5 July 2023). https://eur-lex.europa.eu/legal-content/EN/TXT/?uri=CELEX:52023PC0416.

[74] European Commission (2024c).

issues; 2) it demonstrated strong European solidarity with direct positive effects of its operations to citizens in Europe and beyond; 3) its new tools and activities (e.g. rescEU, ECPP) proved effective to strengthen European resilience against risks. The evaluation also acknowledges that important adaptations and promising initiatives can be expected to yield benefits in the future, such as the further development of rescEU capacities, enhanced coordination capacities of the ERCC for cross-sectoral disasters or ongoing work on scenario-building and analysis.

On 16 July 2025, the Commission presented a legislative and financial proposal to reform the Mechanism for 2028–2034.[75] This proposal is based on the 2024 evaluation, key policy documents like the Niinistö Report, the European Preparedness Union Strategy,[76] and input from a public consultation. The proposal focuses on three key areas: (1) creating a Crisis Coordination Hub: this hub will build on the existing ERCC, using an all-hazard approach to manage complex, cross-sector emergencies; (2) strengthening whole-of-government coordination: key stakeholders, including the military, will be more closely aligned through better planning, training, exercises, and early warning systems; and (3) integrating existing health emergency instruments: these will be closely linked to the Civil Protection Mechanism to improve response during health crises. Additional measures proposed include an increase of the budget to 11 billion EUR for 2028–2034, up from some 8 billion EUR in the previous period, as well as adopting a more flexible legislative tool (Regulation) to allow for financial carry-overs between years.

These steps aim to boost the EU's ability to handle future geopolitical, climate-related, and societal challenges. However, to what extent Member States will go along with the proposals and whether the proposed measures will suffice to address the challenges remains to be seen.

Key Sentences

- **The Central Role of the UCPM**—The Union Civil Protection Mechanism (UCPM) is the EU's principal multilateral instrument for preventing, preparing for, and responding to natural and man-made disasters across and beyond Europe.

---

[75] European Commission, Proposal for a Regulation of the European Parliament and of the Council on the Union Civil Protection Mechanism and Union support for health emergency preparedness and response, and repealing Decision No 1313/2013/EU (Union Civil Protection Mechanism), COM (2025) 548 final, 16 July 2025. https://eur-lex.europa.eu/legal-content/EN/TXT/PDF/?uri=COM:2025:548:FIN

[76] Joint Communication to the European Parliament, the European Council, the Council, the European Economic and Social Committee and the Committee of the Regions on the European Preparedness Union Strategy. JOIN (2025) 130 final.

https://webgate.ec.europa.eu/circabc-ewpp/d/d/workspace/SpacesStore/b81316ab-a513-49a1-b520-b6a6e0de6986/file.bin

- **Membership and Scope**—The Mechanism covers 27 EU Member States plus 10 Participating States and operates globally.
- **Covering the Disaster Cycle**—Decision 1313/2013 frames the UCPM across prevention, preparedness, and response, with distinct Commission and Member State roles.
- **Operational Hubs and Tools**—The Emergency Response Coordination Centre (ERCC), Common Emergency Communication and Information System (CECIS), the European Civil Protection Pool (ECPP), and rescEU form the Mechanism's operational backbone. rescEU provides EU-funded reserve capacities (aerial firefighting, medical, CBRN, shelter, logistics, etc.) to fill collective gaps during simultaneous or large-scale crises.
- **Finance and Logistics**—Progressive EU co-financing supports transport, local logistics, and operational costs, incentivising pre-commitments and rapid deployment.
- **Evolving Usage and Scale**—Activations rose from single digits annually to more than one hundred, driven mainly by COVID-19 and the war in Ukraine, which became the largest and most complex UCPM operation.
- **National Diversity**—European national civil protection systems vary widely in structure, centralisation, volunteer and military roles, and activation thresholds, but no single 'best' model emerged.
- **Complementary EU Instruments**—Directives and strategies (flood, Seveso, nuclear safety, forest/biodiversity/soil strategies) complement the UCPM by reducing risks and building resilience. Complementarities with humanitarian aid enhance effectiveness and efficiency of European emergency action in third countries.
- **Future Reform Agenda**—Recent policy developments push for the establishment of a cross-sectorial Crisis Coordination Hub, stronger whole-of-government alignment, integration of health crises, and a larger budget for 2028–2034.

References

Alexander D (2002) From civil defence to civil protection—and back again. Disaster Prev Manag 11(3):209–213

Backman S, Rhinard M (2018) The European Union's capacities for managing crises. J Contingencies Crisis Manag 26:261–271

Ballester J, Quijal-Zamorano M, Méndez Turrubiates RF et al (2023) Heat-related mortality in Europe during the summer of 2022. Nat Med 29:1857–1866

Boin A, Ekengren M, Rhinard M (2013) The European Union as crisis manager: patterns and prospects. Cambridge University Press

Boin A et al (2014) Civil security and the European Union: a survey of European civil security systems and the role of the EU in building shared crisis management capacities. The Swedish Institute of International Affairs. https://repository.essex.ac.uk/13839/1/102569.pdf

Brazova VK (2021) Stakeholder involvement in the civil security systems: how (dis)similar are the CEE countries compared to the rest of Europe. NISPACEE J Public Admin Policy 14(1):37–59

Brugueras S, Fernandez-Martínez B, Martínez-de la Puente J, Figuerolac J, Montalvo Porro T, Rius C, Larrauri A, Gomez-Barroso D (2020) Environmental drivers, climate change and emergent diseases transmitted by mosquitoes and their vectors in southern Europe: a systematic review. Environ Res 191(Dec 2020):1–17

Bündnis Entwicklung Hilft/IFHV (2025) WordRiskReport 2025. Bündnis Entwicklung Hilft, Berlin. https://weltrisikobericht.de/worldriskreport/#downloads

Cantergiani C et al (2021) ESPON-TITAN. Territorial impacts of natural disasters. Final report. Applied research. https://archive.espon.eu/sites/default/files/attachments/ESPON-TITAN_Main%20Report.pdf

Christensen T, Andreas Danielsen O, LÆhgreid P, Rykka L (2016) Comparing coordination structures for crisis management in six countries. Public Admin 94:316–332

Commission of the European Communities (1987) Cooperation in the community on civil defence. COM (87) 161 final. https://aei.pitt.edu/11966/1/11966.pdf

CRED—Centre for Research on the Epidemiology of Disasters (2020) CRED Crunch no. 60. Technological disasters. https://www.cred.be/sites/default/files/CC60.pdf

CRED—Centre for Research on the Epidemiology of Disasters (2023) 2022 Disasters in numbers. CRED, Brussels. https://www.cred.be/sites/default/files/2022_EMDAT_report.pdf

Dadoudi A (2017) Comparative analysis of civil protection systems. Master's thesis. Aristotle University of Thessaloniki. Department of Economics. https://ikee.lib.auth.gr/record/288791/files/GRI-2017-19077.pdf

Dilley M, Chen RS, Deichmann UK, Lerner-Lam AL, Arnold M, Agwe JN, Buys P, Kjekstad O, Lyon B, Yetman G (2005) Natural disaster hotspots: a global risk analysis. World Bank. https://documents1.worldbank.org/curated/en/621711468175150317/pdf/344230PAPER0Na101official0use0only1.pdf

ECHO—Directorate-General for European Civil Protection and Humanitarian Aid Operations (2025) The national disaster management system. https://civil-protection-humanitarian-aid.ec.europa.eu/what/civil-protection/national-disaster-management-system_en

EEA—European Environment Agency (2022) Climate change as a threat to health and well-being in Europe: focus on heat and infectious diseases EEA Report 7/2022. https://www.eea.europa.eu/en/analysis/publications/climate-change-impacts-on-health

Esposito M (2017) Civil protection systems of the Adriatic and Ionian countries. A review. Adriatic Governance Operational Plan. https://www.adriaticionianeuroregion.eu/wp-content/uploads/2017/05/Adrigov_VI_OICS_Civil-Protection.pdf

European Commission (2012) Nuclear stress tests: confirmation of high safety standards but need for further improvement. Press release 04 October 2012. https://ec.europa.eu/commission/presscorner/detail/en/ip_12_1051

European Commission (2017) Communication from the Commission to the European Parliament, the Council and the Committee of the Regions: strengthening EU Disaster Management: rescEU. Solidarity with responsibility. COM (2017) 773 final. https://eur-lex.europa.eu/legal-content/EN/TXT/PDF/?uri=CELEX:52017DC0773

European Commission (2020) Communication from the Commission to the European Parliament, the Council, the European Economic and Social Committee and the Committee of the Regions: EU Biodiversity Strategy for 2030—Bringing nature back into our lives. COM/2020/380 final. https://eur-lex.europa.eu/legal-content/EN/TXT/?uri=CELEX:52020DC0380

European Commission (2021a) Report from the Commission to the European Parliament and the Council on the Implementation and efficient functioning of Directive 2012/18/EU on the control of major-accident hazards involving dangerous substances for the period 2015–2018. COM (2021) 599 final. https://eur-lex.europa.eu/legal-content/EN/TXT/PDF/?uri=CELEX:52021DC0599
European Commission (2021b) Communication from the Commission to the European Parliament, the Council, the European Economic and Social Committee and the Committee of the Regions: forging a climate-resilient Europe—the new EU Strategy on Adaptation to Climate Change. COM/2021/82 final. https://eur-lex.europa.eu/legal-content/EN/TXT/?uri=CELEX:52021DC0082
European Commission (2021c) Communication from the Commission to the European Parliament, the Council, the European Economic and Social Committee and the Committee of the Regions: new EU Forest Strategy for 2030. COM/2021/572 final. https://eur-lex.europa.eu/legal-content/EN/TXT/?uri=CELEX:52021DC0572
European Commission (2021d) Communication from the Commission to the European Parliament, the Council, the European Economic and Social Committee and the Committee of the Regions: EU Soil Strategy for 2030—Reaping the benefits of healthy soils for people, food, nature and climate. COM/2021/699 final. https://eur-lex.europa.eu/legal-content/EN/TXT/?uri=CELEX:52021DC0699
European Commission (2024a) Report from the Commission to the European Parliament and the Council on progress on implementation of article 6 of the Union Civil Protection Mechanism (Decision No 1313/2013/EU) preventing and managing disaster risk in Europe. COM (2024) 130 final. https://eur-lex.europa.eu/legal-content/EN/TXT/PDF/?uri=COM:2024:130:FIN
European Commission (2024b) Commission Staff Working Document. Evaluation of the Union Civil Protection Mechanism (2017–2022). SWD (2024) 213 final. https://ec.europa.eu/echo/files/evaluation/2024/swd_2024_212_f1_swd_evaluation_en_v4_p1_3451214.pdf
European Commission (2024c) Communication from the Commission to the Council and the European Parliament on the evaluation of the Union Civil Protection Mechanism—Strengthening EU's emergency preparedness. COM (2024) 212 final. https://ec.europa.eu/echo/files/evaluation/2024/com_2024_212_f1_communication_from_commission_to_inst_en_v3_p1_3425414.pdf
European Commission (2025) Report from the Commission to the Council and the European Parliament on the implementation of the Water Framework Directive (2000/60/EC) and the Floods Directive (2007/60/EC). COM (2025) 2 final. https://eur-lex.europa.eu/legal-content/EN/TXT/PDF/?uri=CELEX:52025DC0002
European Commission, Directorate-General for Environment (2000) Vade-mecum of civil protection in the European Union. Publications Office. https://op.europa.eu/en/publication-detail/-/publication/6a0b0c66-757a-42d4-a2c2-68900aec6091
European Commission, Joint Research Centre (2024) INFORM report 2024: 10 years of INFORM: shared evidence for managing crises and disasters. Publications Office of the European Union. https://data.europa.eu/doi/10.2760/555548
Fonseca JD (2005) 1755—O Terramoto de Lisboa. The Lisbon Earthquake. Argumentum
Gordeeva E, Weber N, Wolfslehner B (2022) The new EU Forest strategy for 2030—An analysis of major interests. Forests 13, 1503. https://www.mdpi.com/1999-4907/13/9/1503
Intergovernmental Panel on Climate Change, IPCC (2023) Summary for policymakers. In: Climate Change 2023: synthesis report. Contribution of Working Groups I, II and III to the Sixth Assessment Report of the Intergovernmental Panel on Climate Change. IPCC, Geneva. https://www.ipcc.ch/report/ar6/syr/downloads/report/IPCC_AR6_SYR_SPM.pdf

Kalogiannidis S, Kalfas D, Papaevangelou O, Chatzitheodoridis F (2024) Effectiveness to the emergency management in public organizations: a paradigm from a European Civil Protection Mechanism. J Risk Anal Crisis Response 14(3):291–313. https://jracr.com/index.php/jracr/article/view/501

Klein J, Valkama M, Staudt M, Schmidt-Thomé P, Kallio H (2024) ESPON-TITAN: territorial patterns of natural hazards in Europe. Nat Hazards. https://www.researchgate.net/publication/377561057_ESPON-TITAN_territorial_patterns_of_natural_hazards_in_Europe

Kuipers S, Boin A, Bossong R, Hegemann H (2015) Building joint crisis management capacity? Comparing civil security systems in 22 European countries. Risk, Hazards Crisis Public Policy 6. https://www.researchgate.net/publication/281145766_Building_Joint_Crisis_Management_Capacity_Comparing_Civil_Security_Systems_in_22_European_Countries

Laakso K (2019) Organising civil preparedness in Finland, Sweden and Norway. Comparing national approaches. Master thesis. Faculty of Management and Business. Tampere University

Laurent A, Pey A, Gurtel P, Fabiano B (2021) A critical perspective on the implementation of the EU Council Seveso Directives in France, Germany. Italy and Spain Process Saf Environ Prot 148(April 2021):47–74. https://www.sciencedirect.com/science/article/pii/S0957582020317821

Lettieri E, Masella C, Radaelli G (2009) Disaster management: findings from a systematic review. Disaster Prev Manag 18(2):117–136. https://www.researchgate.net/publication/235313228_Disaster_management_Findings_from_a_systematic_review

López-Carresi A (2012) Emergency management principles. In: Wisner B, Gaillard JC, Kelman I (eds) The Routledge handbook of hazards and disaster risk reduction. Routledge, pp 505–517

Mazhin SA, Farrokhi M, Noroozi M, Roudini J, Hosseini SA, Motlagh ME, Kolivand P, Khankeh H (2021) Worldwide disaster loss and damage databases: a systematic review. J Edu Health Promot:10–13. https://pubmed.ncbi.nlm.nih.gov/34761015/

Morsut C (2014) The EU'S Community Mechanism for Civil Protection: analyzing its development. J Contingencies Crisis Manag 22(3):143–149

Morsut C, Kruke BI (2020) Europeanisation of civil protection: the cases of Italy and Norway. Int Public Manage Rev 20(1):3–23. https://ipmr.net/index.php/ipmr/article/view/361

Munich Re (2025) Data on natural disasters since 1980. Munich Re's Nat Cat Service. https://www.munichre.com/en/solutions/for-industry-clients/natcatservice.html

Navarro D, Cantergiani C, Abajo B, Gomez de Salazar I, Feliu E (2023) Territorial vulnerability to natural hazards in Europe: a composite indicator analysis and relation to economic impacts. Nat Hazards. https://doi.org/10.1007/s11069-023-06165-w

Niinistö, S (2024) Safer together: a path towards a fully prepared union—Strengthening Europe's civil and military preparedness and readiness: Report by Sauli Niinistö, Former President of the Republic of Finland in his capacity as Special Adviser to the President of the European Commission. https://commission.europa.eu/topics/defence/safer-together-path-towards-fully-prepared-union_en

Parker C, Persson T, Widmalm S (2019) The effectiveness of national and EU-level civil protection systems: evidence from 17 member states. J Eur Public Policy 26(9):1312–1334

Peduzzi P, Dao H, Herold C, Mouton F (2009) Assessing global exposure and vulnerability towards natural hazards: the disaster risk index. Nat Hazards Earth Syst Sci 9:1149–1159

Petersson E (2023) Explaining the EU member states' varied positions on European Integration of Civil Protection. A longitudinal comparative analysis of Sweden and Finland. Master thesis. Department of Political Science, University of Gothenburg. https://gupea.ub.gu.se/handle/2077/77281

Prerna J, Pasman HJ, Waldram SP, Rogers WJ, Mannan MS (2017) Did we learn about risk control since Seveso? Yes, we surely did, but is it enough? An historical brief and problem analysis.

J. Loss Prev. Process Ind 49(A), Sept 2016:5–17. https://www.sciencedirect.com/science/article/abs/pii/S0950423016302674

Prior T, Herzog M, Kaderli T, Roth F (2016) International civil protection: adapting to new challenges. ETH Zürich. Risk and Resilience Research Team, Center for Security Studies, ETH Zurich. On behalf of the Federal Office for Civil Protection (FOCP)

Ritchie H, Rosado P, Roser M (2022) "Natural disasters" published online at OurWorldinData.org. https://ourworldindata.org/natural-disasters'

Schmidt-Thomé P (ed) (2006a) The spatial effects and management of natural and technological hazards in Europe—ESPON 1.3.1, Executive Summary Geological Survey of Finland https://archive.espon.eu/sites/default/files/attachments/fr-1.3.1_revised-full.pdf

Schmidt-Thomé P (ed) (2006b) Natural and technological hazards and risks affecting the spatial development of European regions. Espoo, Geol. Survey Finland, Special Paper 42. https://www.researchgate.net/publication/292658616_Natural_and_technological_hazard_maps_of_Europe

Sigala IF, Wakolbinger T (2019) Outsourcing of humanitarian logistics to commercial logistics service providers. An empirical investigation. J Humanit Logist Supply Chain Manag (9)1:47–69. https://www.emerald.com/insight/content/doi/10.1108/jhlscm-12-2017-0073/full/html

Spassov P, Tagarev T (2015) Lithuania. Capabilities, organisations, policies, and legislation in crisis management and disaster response IT4Sec reports 126, March 2015. https://procon.bg/system/files/driver_copl_lithuania_0.pdf

Subedi B, Poudel A, Aryal S (2023) The impact of climate change on insect pest biology and ecology: implications for pest management strategies, crop production, and food security. J Agric Food Res 14:100733. https://www.sciencedirect.com/science/article/pii/S2666154323002405

Talay ÖS (2020) The prospects of Europeanization in the field of civil protection: the cases of Italy and Croatia. Thesis submitted to the Graduate School of Social Sciences of the Middle East Technical University. https://etd.lib.metu.edu.tr/upload/12624855/index.pdf

Treaty on European Union (1992) OJ C191, vol 35. https://eur-lex.europa.eu/legal-content/EN/TXT/PDF/?uri=OJ:C:1992:191:FULL

Treaty on the Functioning of the European Union (2012) OJ C:326/47. https://eur-lex.europa.eu/LexUriServ/LexUriServ.do?uri=CELEX:12012E/TXT:en:PDF

United Nations Office for Disaster Risk Reduction, UNDRR (2017). The Sendai framework terminology on disaster risk reduction. "Disaster". https://www.undrr.org/terminology/disaster

UNDRR (2023). The report of the midterm review of the implementation of the Sendai framework for disaster risk reduction 2015–2030. https://sendaiframework-mtr.undrr.org/media/86858

Villani S, (2017) The EU Civil Protection Mechanism: instrument of response in the event of a disaster. http://revistarue.eu/RUE/052017.pdf

Welle T, Birkmann J (2015) The world risk index—An approach to assess risk and vulnerability on a global scale. J Extreme Events 2(1)

Wood GD (2014) Tambora—The eruption that changed the world. Princeton UP, Princeton

Zweglinski T, Cordero MR (2019) Spanish civil protection system as a part of European Union emergency response. Internal Security 11(2):165–177. https://www.researchgate.net/publication/339311569_Spanish_Civil_Protection_System_as_a_Part_of_European_Union_Emergency_Response

Open Access This chapter is licensed under the terms of the Creative Commons Attribution 4.0 International License (http://creativecommons.org/licenses/by/4.0/), which permits use, sharing, adaptation, distribution and reproduction in any medium or format, as long as you give appropriate credit to the original author(s) and the source, provide a link to the Creative Commons license and indicate if changes were made.

The images or other third party material in this chapter are included in the chapter's Creative Commons license, unless indicated otherwise in a credit line to the material. If material is not included in the chapter's Creative Commons license and your intended use is not permitted by statutory regulation or exceeds the permitted use, you will need to obtain permission directly from the copyright holder.

The International Legal Framework for Disaster Response and Recovery Activities

4

Emanuele Sommario

Abstract

This chapter offers a comprehensive analysis of the international legal framework governing disaster response and recovery activities, identifying existing instruments and highlighting current normative gaps. It begins by clarifying key concepts such as disaster, disaster response, and disaster recovery, noting definitional variations across treaties and soft law instruments, as well as the importance of thresholds, state capacity, and temporal factors in triggering applicable legal regimes. The analysis then turns to the typical provisions found in disaster response frameworks, focusing on the primary role of the affected state; mechanisms for information sharing; the entry and transit of personnel and goods; and the treatment of relief personnel, including questions of privileges, immunities, and recognition of professional titles. It also examines cost-sharing arrangements and liability issues for damage that may be caused by foreign first responders. The chapter underscores how sovereignty remains a central principle in disaster management, with assistance invariably contingent upon the consent of the affected state. The second part of the chapter addresses the less developed field of disaster recovery, focusing on non-binding instruments such as UNGA Resolution 46/182, the Sendai Framework for Disaster Risk Reduction, and the Sustainable Development Goals. The chapter concludes by discussing the potential for a comprehensive treaty on disaster risk management, noting the UN General Assembly's 2024 decision to initiate negotiations based on the recent work of the International Law Commission. While

E. Sommario (✉)
Scuola Superiore Sant'Anna, Pisa, Italy
e-mail: emanuele.sommario@santannapisa.it

© The Author(s) 2026

E. Sommario and P. Billing (eds.), *International Legal and Institutional Frameworks for Disaster Risk Management*, Civil Protection, Disaster Management, and Resilience 4,
https://doi.org/10.1007/978-3-032-11806-6_4

acknowledging legal and political hurdles, the chapter highlights the growing necessity of a unified, binding framework in light of the increasing frequency and impact of disasters worldwide.

Keywords

International disaster law • Response • Recovery • Sovereignty • Cross-border assistance

4.1 Introduction

As evidenced in the Introduction, the increasing frequency and intensity of disasters, whether natural or human-made, have underscored the necessity for a robust international legal framework to guide Disaster Risk Management (DRM) activities throughout the disaster cycle. Recent events have shown that disasters can significantly disrupt social, economic, and political structures, leading to dire humanitarian crises. The international community has recognised the need for a structured response to such crises, resulting in the development of various legal instruments and frameworks aimed at enhancing disaster management. The present chapter will address the norms pertaining to disaster response and recovery activities, while the following chapter will address the remaining phases of the disaster cycle.[1] In particular, we will here focus on the definitions of key terms (Sect. 4.2), the typical contents of disaster response instruments (Sect. 4.3), a description of the (still under-developed) framework concerning recovery activities (Sect. 4.4), and ways forward in enhancing these frameworks through the possible adoption of a new, comprehensive treaty on DRM (Conclusions).

4.2 Defining the Terms *Disaster, Disaster Response,* and *Disaster Recovery*

To establish a comprehensive understanding of the international legal framework for disaster response and recovery, it is crucial to define the key terms involved. Starting with the term 'disaster', different treaties contain different definitions and only become operative when such a definition is met.[2] Eduardo Valencia Ospina—Special Rapporteur of the International Law Commission (ILC) on the topic of the protection of persons in the event of disasters—reminds us that 'disaster is not a term of art and, as such, lacks one single accepted definition'.[3] However, definitions are of essential importance as they are

[1] See the Introduction to this volume for a description of the disaster management cycle.

[2] For a general overview see Bartolini (2018), pp. 10–16.

[3] ILC, Second Report on the Protection of Persons in the Event of Disasters by Mr. Eduardo Valencia-Ospina, Special Rapporteur (7 May 2009), UN Doc A/CN.4/615, p. 193, para 31.

key in identifying the situations in which specific treaties may be invoked, as well as the circumstances under which they cease to be applicable. A definition should also fix reasonable limits on the scope of the subject matter at stake, excluding events such as armed conflict or minor accidents. For the purpose of the present chapter, we will conduct a survey of the definitions contained in legally binding instruments (both bilateral and multilateral) as well as in soft law documents of recognised importance in the area of disaster management.

4.2.1 Disaster

While there is no uniform approach on definitional issues, the large majority of international instruments addressing disasters typically include a definition of this term through certain descriptive elements and general criteria. The elements normally considered concern: (a) the origin of the hazard that might prompt the disaster (natural, human-made, etc.); (b) the consequences of the hazard's impact, i.e. the scale of the damage that must be caused; (c) the values and assets that deserve protection (human life, livelihood, cultural property, natural environment, etc.); (d) the actual or imminent nature of the threat, with some treaties only addressing hazards that have already materialised, and others including incumbent hazards; (e) the requirement that the affected State cannot cope with the consequences of the disaster by using only the human and material resources at its disposal. Most of the time, these elements are combined in the definitions provided. What follows are examples highlighting some of these elements.

We mentioned earlier that 'there is no such thing as a natural disaster',[4] as it is very difficult to draw clear distinctions between natural and human-made disasters. These phenomena are always the result of a complex mix of factors and no single element can be identified as decisive in the start and expansion of a calamitous event. Yet many treaties maintain said distinction in setting out their material scope of application. A good example is the Agreement between Member States and Associate Members of the Association of Caribbean States for Regional Cooperation on Natural Disasters, which defines its scope of application *ratione materiae* as including 'damage caused by any natural phenomenon (hurricane, tornado, storm, tidal wave, flood, tsunami, earthquake, volcanic eruption, landslide, forest fire, epidemic, epizootics, agricultural plague and drought, among others)'.[5] On the other hand, there is no scarcity of treaties that exclusively address certain human-made disasters. For instance, the Convention on Early Notification of a Nuclear Accident is only applicable 'in the event of any accident involving facilities or activities of a State Party or of persons or legal entities under its jurisdiction or control (…) from which a

[4] Hartman and Squires (2006).

[5] Agreement Between Member States and Associate Members of the Association of Caribbean States for Regional Cooperation on Natural Disasters (Caribbean Agreement), 17 April 1999, Art. 1.1.

release of radioactive material occurs or is likely to occur'.[6] Indeed, this treaty avoids providing a definition of *disaster* altogether, and instead establishes a duty of cooperation in case of a nuclear accident, which is deemed to be a disaster in and of itself.

Many definitions make specific mention of the seriousness of the hazardous event, which needs to achieve a certain threshold of destructiveness. For instance, the Agreement on the Establishment of the Civil-Military Emergency Planning Council for Southeastern Europe requires that a disaster causes destruction with the capability to 'seriously endanger the public health, safety and welfare of populations'.[7] Other instruments clearly define the type of destruction or damage that needs to occur to trigger their application, which may involve damage to different types of goods or assets. For instance, according to the Agreement among the Governments of the Participating States of the Black Sea Economic Cooperation (BSEC) on Collaboration in Emergency Assistance and Emergency Response to Natural and Man-Made Disasters, a disaster is an event 'which may cause or have caused significant *physical, social, economic and cultural damage to human lives or environment*'.[8]

While certain definitions assume that the disaster has already impacted a given community, others leave room for preventive action, as they allow measures to be taken even when the calamity is only imminent. For instance, a mutual assistance agreement between Denmark and Germany becomes applicable when the event 'harms *or threatens* the life or health of a large number of people'.[9] Lastly, numerous definitions include a gravity component in that they require that the national response capacities of the affected State are overwhelmed. The above-mentioned definition of the Caribbean Agreement states that the damage caused by the natural hazard must be such that it goes 'beyond the local response capacity' and requires 'regional assistance'.[10]

It must be highlighted that some treaties omit the definitional conundrum altogether, leaving it to the State parties to decide if the situation warrants the application of the agreement. The Inter-American Convention to Facilitate Disaster Assistance, for instance, establishes that it 'shall apply whenever a state party furnishes assistance in response to a request from another state party, except as they otherwise agree'.[11] Finally, several

[6] Convention on Early Notification of a Nuclear Accident, 26 September 1986, Art. 1.1.

[7] Agreement on the Establishment of the Civil-Military Emergency Planning Council for Southeastern Europe, 3 April 2001, Art. II.

[8] Agreement among the Governments of the Participating States of the Black Sea Economic Cooperation (BSEC) on Collaboration in Emergency Assistance and Emergency Response to Natural and Man-Made Disasters, 15 April 1998, Art. 2 (emphasis added).

[9] Agreement on Mutual Assistance in the Event of Disasters or Serious Accidents between Denmark and the Federal Republic of Germany, 16 May 1985, Art. 1.1 (emphasis added).

[10] Caribbean Agreement, Art. 1.

[11] Inter-American Convention to Facilitate Disaster Assistance, 7 June 1991, Art. I.a. This deference to the parties' will is made even clearer in other instruments, which explicitly subordinate the application of the treaty to a determination made by one of them that an event amount to a 'disaster'. For instance, the Agreement between the Argentine Republic and the Republic of Chile on Cooperation

instruments explicitly exclude armed conflicts and other situations of violence from the legal definition of 'disaster'. The Guidelines for the Domestic Facilitation and Regulation of International Disaster Relief and Initial Recovery Assistance developed by the International Federation of the Red Cross and Red Crescent Societies (IFRC), for example, make it clear that their definition does not include armed conflict situations.[12] The reasons behind this choice are due to the different operational challenges posed by armed violence situations, as well as by the fact that armed conflicts are already extensively regulated—including with respect to humanitarian activities—by a specific branch of international law, namely, International Humanitarian Law (IHL), which prevails over other legal regimes as *lex specialis*.[13]

For the sake of completeness, it is worth mentioning the definition of disaster developed by the ILC in its Draft Articles (DAs) on the Protection of Persons in the Event of Disaster. According to DA 3(a), 'disaster' means 'a calamitous event or series of events resulting in widespread loss of life, great human suffering and distress, mass displacement, or large-scale material or environmental damage, thereby seriously disrupting the functioning of society'.[14] It is interesting to note that the definition adopts a holistic approach as it includes both natural and human-made hazards. Furthermore, the reference to a 'calamitous event' serves to establish a threshold whereby only extreme events are covered.[15] According to the ILC Commentary, the definition equally covers 'sudden-onset events (such as an earthquake or tsunami) and […] slow-onset events (such as drought or sea-level rise), as well as frequent small-scale events (floods or landslides)'.[16] Based on the above survey it can be concluded that there is no single definition of the term, and that the applicability of each instrument must be assessed on the basis of the specific terms employed in its provisions.[17]

---

in Case of Disasters, adopted on 8 August 1997, states that the term *disaster* 'shall mean any event so described by the Party requiring assistance', Art. 1.

[12] IFRC, Guidelines for the Domestic Facilitation and Regulation of International Disaster Relief and Initial Recovery Assistance (2007) (IDRL Guidelines), 2.1.

[13] ILC, Draft articles on the protection of persons in the event of disasters, with commentaries, UN Doc A/71/10, p. 46, para. 9. On the legal regime applicable to disasters occurring in areas affected by armed conflicts, see Zorzi Giustiniani (2021), pp. 39–49.

[14] ILC, Draft articles on the protection of persons in the event of disasters, p. 5.

[15] Ibid., p. 6.

[16] Ibid.

[17] Significantly, one important author reminds us that there is no international consensus 'on how great a catastrophe has to be in order to be considered a disaster for legal purposes, nor is there any agreement on what criteria should be used to measure its scale', Focarelli (2021), para. 8.

4.2.2 Disaster Response

Moving to 'disaster response', it is commonly considered to encompass activities focused on the immediate and short-term needs of disaster-affected communities. However, even in this case, there is no scarcity of legal definitions. The Caribbean Agreement defines the term as involving 'the activities carried out immediately following the disaster, and includes, inter alia, rescue and response activities, the provision of health services, food, shelter, water, sanitation facilities and other basic necessities for survival'.[18] The UN Report on Disaster Risk Reduction (DRR) terminology describes it as '[a]ctions taken directly before, during or immediately after a disaster in order to save lives, reduce health impacts, ensure public safety and meet the basic subsistence needs of the people affected', also including in the definition measures taken shortly before the disaster strikes aimed at mitigating its effects.[19] Other instruments use different terms to indicate the same notion. The IDRL Guidelines, for instance, choose the term 'disaster relief' to indicate '[g]oods and services provided to meet the immediate needs of disaster-affected communities'.[20] Furthermore, what is not entirely clear-cut is the division between disaster response and the subsequent recovery phase. Indeed, some response actions, such as the provision of temporary housing and water supplies, 'may extend well into the recovery stage'.[21]

4.2.3 Disaster Recovery

Looking at the notion of 'disaster recovery', even this term lacks a universally agreed definition. The UN Expert Working Group on DRR defines 'recovery' as '[t]he restoring or improving of livelihoods and health, as well as economic, physical, social, cultural and environmental assets, systems and activities, of a disaster-affected community or society, aligning with the principles of sustainable development and 'build back better', to avoid or reduce future disaster risk'.[22] Other definitions appear to identify different stages in the recovery phase, depending on the chronological proximity to the disastrous event. The

---

[18] Caribbean Agreement, Art. 1.10.

[19] UN General Assembly (UNGA), Report of the Open-Ended Intergovernmental Expert Working Group on Indicators and Terminology Relating to Disaster Risk Reduction (1 December 2016) UN Doc A/71/644, p. 22.

[20] IFRC, IDRL Guidelines, 2.2.

[21] UN International Strategy for Disaster Reduction (ISDR), UNISDR Terminology on Disaster Risk Reduction (31 December 2009) p. 25.

[22] UNGA, Report of the Open-Ended Intergovernmental Expert Working Group, p. 21. The concept of 'build back better' is described in the same document as '[t]he use of the recovery, rehabilitation and reconstruction phases after a disaster to increase the resilience of nations and communities through integrating disaster risk reduction measures into the restoration of physical infrastructure and societal systems, and into the revitalization of livelihoods, economies and the environment', ibid., p. 11.

IDRL Guidelines, for example, mention the notion of 'initial recovery assistance', which refers to the 'means, goods and services intended to restore or improve the pre-disaster living conditions of disaster-affected communities, including initiatives to increase resilience and reduce risk, provided for an initial period of time, as determined by the affected State, *after the immediate needs of disaster-affected communities have been met*'.[23] Similarly, the Cluster Working Group on Early Recovery introduces the notion of 'early recovery', defined as 'a multidimensional process of recovery that begins in a humanitarian setting' which, guided by development principles, seeks to 'catalyse sustainable development opportunities'.[24] This component of recovery is sometimes also referred to as 'rehabilitation' and overlaps with disaster response.[25] It is followed by measures aimed at fully rebuilding and restoring functionality across all sectors and aspects of society, a phase that may be referred to as 'reconstruction' or 'long-term recovery'.

While no unified definition of the term 'recovery' has emerged, it is reasonable to maintain that recovery activities begin soon after the emergency phase has ended and aim at restoring or improving livelihoods and health, as well as the economic, physical, social, cultural, and environmental assets, systems, and activities of a disaster-affected community. Moreover, they should ideally be geared towards reducing the risk of future disasters, and hence linked to the prevention, preparation, and mitigation agendas.

4.3 Typical Contents of a Disaster Response Law Instrument

As mentioned, although no flagship treaty has been established so far to regulate disaster response, international law has not remained silent on the matter. In fact, the international legal framework on disaster relief has evolved on multiple fronts at the global and regional levels, spanning various sectors of international law.[26] Additionally, numerous *soft law* instruments, including resolutions, declarations, codes, models, and guidelines, have emerged. While these are not legally binding, they still carry significant weight, reflecting international consensus and best practices. Numerous issues pertaining to disaster response are regulated by international disaster law instruments. While some are specific to particular agreements, some issues are a recurrent feature and will be briefly presented in the next paragraphs. Our analysis will follow the successive phases of a typical cross-border disaster response mission.

[23] IFRC, IDRL Guidelines, 2.3 (emphasis added).

[24] Cluster Working Group on Early Recovery, Guidance note on Early Recovery, April 2008, p. 6. Note that the document concerns both disaster and armed conflict situations.

[25] IFRC, Disaster Risk Governance Guidelines. Strengthening Laws, Policies and Plans for Comprehensive Disaster Risk Management, 2024, p. 49.

[26] de Guttry (2012), p. 3 ff.

4.3.1 The Primary Role of the Affected State

In determining the appropriate response of an affected State to a disaster event, it is important to stress that even in such circumstances international action is constrained by the core principles of State sovereignty and non-intervention. The International Court of Justice (ICJ) describes mutual respect for territorial sovereignty between States as an essential pillar of international relations.[27] UN General Assembly (GA) Resolution 46/182 affirms that in the context of disaster response, 'the sovereignty, territorial integrity and national unity of States must be fully respected in accordance with the Charter of the United Nations'.[28] Hence, the initiation and implementation of international relief assistance are contingent upon the consent of the affected State, whose authorities have the primary responsibility for assisting the victims of disasters that occur within their territory. This means that national authorities have the primary role in the direction, control, coordination, and supervision of the provision of disaster assistance. The primacy of the affected State derives both from its sovereign prerogatives and from its responsibility towards the affected population within its territory.[29] The idea that the affected State retains control over all aspects of the provision and distribution of assistance on its territory, including determining the moment of commencement and termination of the assistance, is explicitly enshrined in several treaties. The ASEAN Agreement on Disaster Management, for example, stipulates that '[t]he Requesting or Receiving Party shall exercise the overall direction, control, co-ordination and supervision of the assistance within its territory'.[30]

4.3.2 Early Phases of the Disaster: Requests and Offers of Assistance

If a disaster has materialised or is imminent, the affected State may issue a request for assistance. Virtually all disaster law instruments include specific provisions on the topic, which may include details on which authority within the State is entitled to issue the request, to whom, in what form, and on the scope of the assistance required. For instance, the Tampere Convention on the provision of telecommunication assistance in disaster settings stipulates that '[a] State Party requiring telecommunication assistance for disaster mitigation and relief may request such assistance from any other State Party, either directly

---

27 ICJ, Corfu Channel case (United Kingdom of Great Britain and Northern Ireland v. Albania), Judgment of 9 April 1949, I.C.J. Reports 1949, p. 4.

28 UNGA, Strengthening of the coordination of humanitarian emergency assistance of the United Nations, UN Doc. A/RES/46/182 (19 December 1991) para 3.

29 Institut de Droit International, Humanitarian assistance, Bruges Session (2 September 2003), Art. III.1.

30 ASEAN Agreement on Disaster Management and Emergency Response, 26 July 2005, Art. 3.2.

or through the operational coordinator'.[31] Before issuing its requests, the affected State should promptly assess existing needs. In doing so, consideration should be given to undertaking joint needs assessments with the United Nations, the EU, and other assisting humanitarian organisations.[32]

A request made under a specific treaty serves to trigger the procedures and processes established under that agreement. Given the number of existing agreements, it is possible that multiple such instruments, at the global, regional, and bilateral levels, would potentially regulate the provision of assistance with respect to a particular disaster.[33] However, under general international law, receiving a request for assistance does not automatically oblige the addressees to provide it. Obligations to provide assistance, upon request, may instead be found in some specific treaties. Art. 222 of the Treaty on the Functioning of the EU, for instance, imposes an obligation to assist on both the Union and its member States, if the political authorities of a disaster-affected Member State so request.[34] But in the absence of specific treaty obligations, there is no legal duty of solidarity vis-à-vis the State struck by a calamity. According to the ILC's DAs on the protection of persons in the event of disaster, the addressee of a request for assistance is only required to 'expeditiously give due consideration to the request and inform the affected State of its reply'.[35]

On the other hand, a State affected by a disaster invariably receives offers of assistance from other States, international organisations, and NGOs. The ILC's DA 12 ('Offers of Assistance') establishes that 'in the event of disasters, States, the United Nations and other potential assisting actors may provide assistance to the affected State'. The provision is mainly intended to highlight that such offers cannot be regarded as interference in the affected State's internal affairs.[36] On the other hand, the ILC has expressly maintained that '[s]uch offers (…) are essentially voluntary and should not be construed as a recognition

---

[31] Tampere Convention on the Provision of Telecommunication Resources for Disaster Mitigation and Relief Operations, 18 June 1998, Art. 4.1.

[32] IFRC, IDRL Guidelines, 10.1.

[33] Indeed, many agreements include specific provisions on their relationship with other international agreements. For example, the Tampere Convention establishes that it 'shall not affect the rights and obligations of States Parties deriving from other international agreements or international law', Art. 10.

[34] Treaty on the Functioning of the European Union (TFEU), 13 December 2007, Art. 222. It should be stressed, however, that according to the Declaration on Article 222 of the Treaty on the Functioning of the European Union' (no. 37), annexed to the Final Act of the Lisbon Conference, 'none of the provisions of Article 222 is intended to affect the right of another Member State to choose the most appropriate means to comply with its own solidarity obligation towards that Member State'. In other terms, each Member State, even if under a duty to assist the victim State, will maintain the right to choose, in good faith, the concrete modalities of assistance. See Gestri (2012), p. 114.

[35] ILC, Draft articles on the protection of persons in the event of disasters, Art. 10.2.

[36] ILC, Report of the International Law Commission: Sixty-Eighth Session (2 May–10 June and 4 July–12 August 2016), UN Doc A/71/10, 2016, p. 57, para. 3.

of the existence of a legal duty to assist'.[37] Hence, the ILC confirmed that no legal obligation to provide assistance exists unless provided for by a specific treaty.

4.3.3 Entry in the Territory of a Disaster-Affected State

Once an international assistance mission has gained access to the territory of a disaster-affected State, a series of legal obstacles could arise capable of slowing down operations and jeopardising their effectiveness. To prevent this, several treaties impose obligations to facilitate the speedy provision of foreign assistance. According to the ILC's DA 15 ('Facilitation of external assistance').

[t]he affected State shall take the necessary measures, within its national law, to facilitate the prompt and effective provision of external assistance, in particular regarding:

(a) relief personnel, in fields such as privileges and immunities, visa and entry requirements, work permits, and freedom of movement; and,
(b) equipment and goods, in fields such as customs requirements and tariffs, taxation, transport, and the disposal thereof.

In the following pages, we will survey the solutions identified in different international legal instruments to address some of the issues at hand.

4.3.3.1 Visas

Starting with the issue of visas, it must be stressed that many States have, by bilateral or multilateral agreement, waived certain requirements to gain access to the respective territories for each other's nationals.[38] However, when a particular disaster-affected State falls outside one of these agreements, problems of access can arise for relief providers. While international relief personnel are frequently allowed to enter without a visa or on tourist visas, these temporary arrangements may not prove effective beyond the immediate response phase, when normal rules (e.g. the requirement for work permits) become applicable again.[39] For this reason, international agreements establishing specific obligations in this area are particularly helpful. The Inter-American Convention, for instance, states

[37] Ibid., para 2.

[38] A good example is the Schengen Area, through which EU and associated States have created a system which allows their legal residents to travel across the Area's internal borders without being checked or having to show their passports, https://home-affairs.ec.europa.eu/policies/schengen/schengen-area_en.

[39] For the practical problems faced by humanitarian agencies in these circumstances, see Fisher (2007), pp. 116–117.

that each State party 'shall provide disaster relief personnel with the necessary immigration documents and facilities'.[40] Similarly, the 1984 Draft Convention on Expediting the Delivery of Emergency Assistance provides that the receiving State shall promptly issue, without cost, multiple transit, entry, and exit visas for personnel representing assisting States or organisations.[41] Other treaties ask the requesting State to waive any visa requirement altogether. For instance, the treaty between the Netherlands and Belgium provides that the 'members of an emergency unit shall be exempt from the requirement to carry a valid document for crossing the frontier'.[42]

4.3.3.2 Access of Goods and Equipment

Humanitarian assistance and relief during a disaster may require the affected State to import essential goods and equipment to meet the needs of its population. This phase can present significant challenges that hinder the timely delivery of aid, including customs delays, the absence of exemptions from duties, restrictions, and compliance with standard customs procedures.[43] While a substantial body of law has emerged to smooth these procedures, the identified solutions remain diverse, with variations in scope and content.[44]

The issue of the facilitation and simplification of border formalities, as well as the waiver of taxes and other financial requirements, is addressed in several agreements and soft law documents. Special schemes have been developed to facilitate the clearance of customs procedures and the elimination of tariffs and other barriers for the importation of goods. As far as customs duties are concerned, a number of States apply a distinct legal regime regulating the 'temporary admission'[45] of goods into their jurisdictions and have developed expedited procedures for the import (and export) of relief items. Provision has also typically been made for the complete waiver of import (and export) duties and taxes, as well as for the waiving or easing of non-financial restrictions.

The prompt clearance of humanitarian assistance goods, which is key to ensuring the timely provision of assistance, is a further common requirement. For instance, the 1973

---

[40] Inter-American Convention, Art. VII.a.

[41] Draft Convention on Expediting the Delivery of Emergency Assistance, UN Doc. A/39/267/Add.2, 18 June 1984.

[42] Convention on Mutual Assistance in Combating Disasters and Accidents, Netherlands-Belgium, 14 November 1984, Art. 6.3.

[43] Fisher (2007), p. 98.

[44] For an overview, see Adinolfi (2012).

[45] The term 'temporary admission' is defined as 'the Customs procedure under which certain goods (including means of transport) can be brought into a Customs territory conditionally relieved from payment of import duties and taxes and without application of import prohibitions or restrictions of an economic character; such goods (including means of transport) must be imported for a specific purpose and must be intended for re-exportation within a specified period and without having undergone any change except normal depreciation due to the use made of them', see Convention on Temporary Admission, 26 June 1990, Art. 1(a).

International Convention on the Simplification and Harmonization of Customs Procedures requires the 'clearance of relief consignments for export, transit, temporary admission and import [to] be carried out as a matter of priority',[46] and preferably without regard to the country of origin. Other measures that may be envisaged to facilitate the clearance of goods include allowing simplified procedures, such as the submission of a shortened or incomplete goods declaration,[47] to be completed within a specified timeframe; enabling clearance outside designated business hours or away from customs offices[48]; and providing waivers for the examination of goods.[49]

Numerous bilateral treaties include specific provisions for the exemption of import and re-export restrictions on medical supplies, including narcotic drugs, under certain conditions.[50] These conditions often stipulate that such supplies may only be transported to address an urgent medical need, used exclusively by qualified medical personnel in compliance with the laws of the Contracting State to which the emergency team belongs, or that the head of the assistance team must formally declare the contents and nature of the imported medication.[51] Additionally, the importation of these goods is typically exempt, in mutual relations, from the applicability of international conventions regulating narcotic drugs, insofar as such conventions would otherwise prohibit their importation.

4.3.4 Work Permits and Professional Qualifications

Closely connected to the issuance of visas is the issue of work permits.[52] Assuming that foreign disaster relief personnel gain access to the affected State, they may not have the status necessary to work within its territory. Sometimes this is due to the specific type of visa they enter the country on, which does not allow them to carry out any paid activity. Several instruments recommend that receiving States facilitate or lift the requirement of obtaining work permits for disaster assistance personnel. For instance, the Draft Model

[46] International Convention on the Simplification and Harmonization of Customs Procedures, 18 May 1973, as revised on 26 June 1999, Specific Annex J(5), Art. 5.

[47] Convention on Temporary Admission, Annex A concerning temporary admission papers (ATA Carnets and CPD Carnets), Art. 2.

[48] Recommendation of the Customs Cooperation Council to Expedite the Forwarding of Relief Consignments in the Event of Disasters, document T2–423, 8 June 1970, para. 7.

[49] Convention on Customs Procedures, Specific Annex J(5), Art. 3.

[50] See, for instance, Netherlands-Belgium, Convention on Mutual Assistance in Combating Disasters and Accidents, 14 March 1984, Art. 7.6; or the Convention entre le Gouvernement de la République française et le Gouvernement du Royaume de Belgique sur l'assistance mutuelle en cas de catastrophe ou d'accidents grave, 21 April 1981, Art. 5.5.

[51] Agreement between the Republic of Austria and the Federal Republic of Germany Concerning Mutual Assistance in the Event of Disasters or Serious Accidents, 23 December 1988, Art. 7(4).

[52] For a general overview of the issue, see Silingardi (2012a), pp. 565–568.

Agreement on International Medical and Humanitarian Law provides that 'the Government of (the receiving State) shall provide the personnel of the organization with work permits to this effect. Those permits may also be provided in the form of a stamp affixed in the documents issued by the organization'.[53] In other cases, international instruments contain broad provisions that, although not specifically mentioning work permits, can be interpreted as including their issuance within their scope. The Framework Convention on Civil Defence Assistance, for instance, provides that '[t]he Beneficiary State shall, within the framework of national law, grant all privileges, immunities, and facilities necessary for carrying out the assistance'.[54] In conclusion, a specific work authorisation is an important component of operational activities of international disaster relief personnel in the receiving State, and it should not be assumed that a provision governing the entry of such personnel necessarily permits them to work upon entry.

Even if a receiving State has granted international disaster relief personnel permission to work through a specific permit, there may still be additional regulatory hurdles. Professionals involved in disaster relief, such as doctors, nurses, drivers, and pilots, are typically subject to licensing requirements specific to each country. Several regional agreements have been established to address the recognition of foreign certificates and diplomas, but they rarely take into account the need for expedited recognition, which the disaster situation presents.[55] Some disaster-related instruments have begun to address the issue, asking for a generalised and expedited recognition scheme. For example, the 1984 Draft Convention on Expediting the Delivery of Emergency Assistance established that receiving States would 'recognise university degrees, professional certificates and other certificates of competency and licences held by relief personnel and necessary for the performance of the agreed function'.[56] Moreover, broad provisions asking the requesting State to afford facilities to intervening actors might be interpreted to also include a waiver on the requirement of recognising foreign degrees and other titles.

4.3.5 Transit and Freedom of Movement

For disaster relief efforts to be effective, unrestricted freedom of movement for humanitarian actors should be guaranteed. This overarching principle encompasses several specific

[53] International Law Association, Draft Model Agreement on International Medical and Humanitarian Law, Art. 15(1), reproduced in the Report of the 59th Conference of the International Law Association (Belgrade, 17–23 August 1980).

[54] Framework Convention on Civil Defence Assistance, 22 May 2000, Art. 4.5.

[55] See for instance the Regional Convention on the Recognition of Studies, Diplomas and Degrees in Higher Education in Asia and the Pacific, 16 December 1983, Art. 5.

[56] UNDRO Secretary-General, Proposed Draft Convention on Expediting the Delivery of Emergency Assistance, A/39/267/Add.2-E/1984/96/Add.2, 18 June 1984, annex, Art. 7.2.d.

aspects, including the right to transit through third-party states and unimpeded movement within the affected state.

With respect to the first issue, several instruments specifically address the passage of goods through the territory of States that are not directly affected by a disaster. For example, the above-mentioned Draft Convention on Expediting the Delivery of Emergency Assistance asks States to 'grant the right of transit across or over their territories to the relief consignments, equipment and personnel of the Assisting State or organization, and to their means of transport, proceeding to or returning from the Receiving State'.[57] Other instruments go even further, requiring transit States to facilitate transit across their territory, or to 'ensure all necessary support'.[58]

Freedom of movement within the territory of a receiving State is a critical condition for the effective delivery of disaster assistance. Access challenges do not necessarily end at the national border; an assisting actor may be permitted entry into the receiving State yet be unable to reach the actual disaster-affected area. In some instances, such impediments may stem from deliberate restrictions imposed by the receiving State for political or security reasons. In others, they may arise from practical logistical difficulties, such as destroyed infrastructure or insufficient capacity to support the large-scale transport of relief supplies. In cases involving deliberate restrictions, certain international instruments establish a negative obligation on the receiving State to refrain from unduly obstructing access to the affected area. Conversely, where logistical challenges are at issue, some instruments impose a positive obligation on the receiving State to facilitate transit through appropriate means.

As far as deliberate restrictions are concerned, several soft law instruments suggest the existence of a duty on the receiving State not to unduly limit access to the disaster relief area. For example, the UNITAR Model Rules for Disaster Relief Operations provide that '[t]he receiving State shall permit the designated relief personnel freedom of access to, and freedom of movement within, disaster-stricken areas that are necessary for the performance of their specifically agreed functions'.[59] Similarly, according to the Institute of International Law, the receiving State shall grant humanitarian organisations 'full and free access to all the victims and ensure the freedom of movement and the protection of personnel, goods and services provided'.[60]

---

[57] Ibid., Art. 20.1.a.

[58] See, for example, the Convention on Assistance in the Case of a Nuclear Accident or Radiological Emergency, 26 September 1986, Art. 9. UNGA Resolution 46/182 provides that 'States in proximity to emergencies are urged to participate closely with the affected countries in international efforts, with a view to facilitating, to the extent possible, the transit of humanitarian assistance', Annex, para. 7.

[59] United Nations Institute for Training and Research (UNITAR), Model Rules for Disaster Relief Operations, 1982, Rule 16.

[60] Institute of International Law, Resolution on Humanitarian Assistance, Adopted at the Bruges session on 2 September 2003, Sect. VII, para. 3.

Some provisions concerning freedom of movement require that the authorities of the receiving State undertake positive steps to facilitate the free circulation of relief workers and their goods and equipment. For instance, GA Resolution 46/182 stipulates that the UN Emergency Relief Coordinator has the responsibility of 'actively facilitating, including through negotiation if needed, the access by the operational organizations to emergency areas for the rapid provision of emergency assistance by obtaining the consent of all parties concerned, through modalities such as the establishment of temporary relief corridors where needed, days and zones of tranquillity and other forms'.[61]

4.3.6 Privileges and Immunities of Foreign Relief Workers

Disaster relief operations give rise to significant legal questions concerning the status of both the personnel involved and the organisations to which they belong. Key issues in this regard include the identification of relief personnel, the use of internationally recognised distinctive emblems, and the issuance of identification cards to individual members of disaster response teams. Among the most important issues raised, however, is the extent to which privileges, immunities, and other legal facilities are granted to relief workers. This issue is frequently addressed in international disaster law instruments, and relevant provisions can also be found in broader legal frameworks that are not exclusively focused on natural or human-made disasters.

In fact, several international conventions grant privileges and immunities to specific entities and categories of individuals, including diplomatic and consular representatives of governments,[62] as well as intergovernmental organisations and their personnel.[63] The scope of these privileges and immunities varies depending on the beneficiary but generally encompasses facilitation of entry and work permits; freedom of movement within the host state; and the inviolability of premises, archives, and communications. Additionally, these protections often include exemptions from personal service obligations, customs duties, and taxation, as well as immunity from arrest, detention, and legal proceedings.[64] Furthermore, international organisations are typically endowed with juridical personality under domestic law, granting them the capacity to enter into contracts, acquire and dispose of property, and initiate legal proceedings.

The general rules outlined above may be applicable to the scenarios addressed in the present chapter. For example, international organisations and their personnel involved in disaster relief operations are, in principle, entitled to the privileges and immunities

[61] UNGA Resolution 46/182, Annex, para. 35.d.

[62] See, for instance, the Vienna Convention on Diplomatic Relations, 18 April 1961; and the Vienna Convention on Consular Relations, 24 April 1963.

[63] See the Convention on the Privileges and Immunities of the United Nations, 13 February 1946, and the Convention on the Privileges and Immunities of Specialized Agencies, 21 November 1947.

[64] On the topic of immunities see Shaw (2021).

conferred by the relevant international conventions. However, this raises the question of whether the particular characteristics of disaster relief operations warrant a tailored approach to the privileges and immunities ordinarily recognised under international law. In particular, it is asked if these prerogatives should be afforded to entities or persons involved in disaster relief operations who are normally not covered by them, such as state officials participating in international relief operations, or the foreign personnel of humanitarian NGOs.[65]

Starting with the first category, their privileges and immunities are the object of specific provisions in some multilateral treaties. These provisions usually impose on the receiving State an obligation to grant certain entitlements deemed necessary for the performance of disaster relief work. For example, Article 8 of the Convention on Assistance in the Case of a Nuclear Accident provides that the requesting State is bound to afford to the assisting party immunity from arrest, detention and legal process and exemption from taxation, duties, or other charges for the personnel of the assisting party in respect of the performance of assistance functions. Also included is an exemption from taxation, duties and other charges and immunity from seizure, attachment, or requisition of equipment and property used for the purpose of assistance.[66] Bilateral agreements almost invariably contain similar clauses. For instance, under the agreement between Argentina and Spain, members of emergency teams involved in assistance in the event of disasters 'shall be immune from the administrative, civilian and criminal jurisdiction of the requesting State with respect to activities implemented in the performance of their duties in the territory of that State'.[67]

Moving to the staff of NGOs involved in disaster relief, they are generally not granted privileges and immunities under international law.[68] However, the relevant provisions of some instruments have been interpreted in such a way as to include non-governmental organisations among the beneficiaries of privileges, immunities, and facilities. For instance, according to the Tampere Convention, the requesting States shall, within the limits of its national law, 'afford to persons, other than its nationals, and to organizations, other than those headquartered or domiciled within its territory (…) the necessary privileges, immunities, and facilities for the performance of their proper functions'.[69] The language used would seem to provide for the possibility of extending such rights to NGO personnel, as long as they are engaged in telecommunications assistance.[70]

Against this background, it must be recalled that privileges and immunities are not absolute in nature. Receiving States retain the sovereign right to deny entry or operation

---

65 Fisher (2007), pp. 39–40.

66 Convention on Assistance in the Case of a Nuclear Accident, Art. 8, paras. 2 and 3.

67 Agreement on Cooperation on Disaster Preparedness and Prevention and Mutual Assistance in the Event of Disasters, Spain-Argentina, 3 June 1988 Art. XVIII.

68 Silingardi (2012b), p. 479.

69 Tampere Convention, Art. 5.

70 Fisher (2007), p. 40.

within their territory to an organisation and may declare a specific official or representative persona non grata, thereby excluding them from the territory. Moreover, entities and individuals benefiting from such privileges and immunities are bound by the relevant legal instruments to refrain from their abuse. Finally, the immunities granted to officials may be waived by the sending State or organisation.[71]

4.3.7 Protection of Relief Personnel

The safety and security of humanitarian relief personnel is an important condition for the provision of assistance. Several instruments contain broadly worded provisions on the protection of humanitarian personnel, including the staff of NGOs. For example, some multilateral conventions provide that the receiving State shall ensure the protection of personnel, equipment, and materials brought into its territory by, or on behalf of, the assisting party.[72] Several bilateral treaties stipulate that the authorities of the requesting party must ensure the protection of emergency teams deployed by the assisting party.[73] These provisions, which pertain to the safeguarding of personnel acting on behalf of the assisting State, may also be interpreted to encompass representatives of non-governmental organisations, provided such entities have been formally engaged in the disaster relief effort.

Certain legal instruments include a general clause concerning the protection of disaster relief personnel, and extend this protection to individuals affiliated with NGOs, irrespective of any formal legal connection to the assisting State. This is made possible through the use of a broad definition of the notion of 'assisting personnel', which may encompass NGO staff operating within the framework of the relief effort. For example, the Inter-American Convention obliges the receiving State to make its best efforts to protect the personnel, equipment, and materials brought into its territory as part of a relief operation.[74] It further stipulates that its provisions apply automatically to NGOs that are part of a State or intergovernmental organisation's relief mission.[75] Additionally, the treaty allows for its application even to NGOs operating independently of such missions, provided there is mutual agreement between the receiving State and the organisation.[76]

---

[71] See, for instance, the Convention on the Privileges and Immunities of the United Nations, Art. 20.

[72] See e.g. the Convention on Assistance in the Case of a Nuclear Accident, Art. 3.b.

[73] See for instance the Convention entre le Gouvernement de la République française et le Gouvernement du Royaume de Belgique sur l'assistance mutuelle en cas de catastrophe, Art. 7.3.

[74] Inter-American Convention, Art. IV.c.

[75] Ibid., Art. XVI.b.

[76] Ibid., Art. XVI.c.

4.3.8 Costs Relating to Disaster Response Missions

Provisions addressing the allocation of costs in disaster response operations vary considerably in both structure and substance. Multilateral treaties and resolutions that establish international or regional disaster relief mechanisms often include clauses creating funding arrangements intended to finance relief operations through the mechanism itself.[77] In contrast, bilateral treaties typically approach cost-sharing from a different angle. Rather than establishing pre-existing funding structures, they commonly contain clauses that create a presumption regarding which party will bear the operational costs. These presumptions may be subject to modification based on specific circumstances outlined in the treaty. Notably, there is no uniform practice regarding the default allocation of financial responsibility: while some treaties presume that the receiving State will cover the costs, others place this obligation on the assisting State.[78] Within the framework of the EU Civil Protection Mechanism, the relevant rule provides that '[u]nless agreed otherwise, the requester of assistance shall bear the costs of assistance provided by Member States', although the provision specifies that the assisting State may 'offer its assistance entirely or partially free of charge' and 'also waive all or part of the reimbursement of its costs at any time'.[79]

4.3.9 Liability and Compensation

International disaster response operations entail a number of dangers to response personnel and material due to unfortunate events such as accidents or disease. Moreover, many activities involved in disaster response may carry a high risk of criminal or civil liability, as relief personnel could cause damage to the assets of the receiving State or to private persons present in its territory.[80] Provisions concerning liability and compensation in instruments related to disaster relief address the question of allocation of liability both between States and with respect to individuals. As may be expected, international instruments frequently focus on inter-State allocation of liability and compensation, while only a few instruments address liability of or compensation to individuals.

Under the majority of instruments, liability for any harm caused is based on a predetermined formula. For example, many treaties obligate the receiving State to bear all risks and claims resulting from, occurring in the course of, or otherwise connected with

[77] See, for example, ASEAN Agreement on Disaster Management and Emergency Response, Art. 24; or the Andean Council of Foreign Ministers, Decision No. 529 establishing the Andean Committee for Disaster Prevention and Response (CAPRADE), 7 July 2002, Art. 5.

[78] This is, for instance, what is established by the Agreement Establishing the Caribbean Disaster Emergency Response Agency, 26 February 1991, Art. 19.

[79] Commission Implementing Decision 2014/762/EU of 16 October 2014 laying down rules for the implementation of Decision No 1313/2013/ EU, Art. 39.

[80] See ILC, Memorandum of the Secretariat, Protection of Persons in the Event of Disasters, 31 March 2008, A/CN.4/590, p. 134 ff.

the assistance rendered on its territory. With respect to multilateral conventions, such provisions often include additional clauses to the effect that the receiving State agrees to hold the assisting State or personnel exempted from obligations that could arise in case of claims in connection with the assistance provided, except in respect of liability of individuals having caused damage by wilful misconduct or by gross negligence.[81] In other cases, the receiving State agrees to pay any claims which could be brought by third parties against the assisting State for loss or damage.[82]

If the relief operation is conducted within the framework of a treaty that does not address the issue of the civil and criminal liability of first responders, the rules applicable need to be traced within the national legal framework of the receiving State. As far as members of national civil protection agencies are concerned, it is accepted in customary international law that they could benefit from a certain level of immunity from process in the courts of the receiving State. However, neither the precise extent of that immunity nor the degree to which it would protect governmental relief personnel is entirely clear.[83]

4.3.10 Additional Relevant Rules

While disaster response law instruments may provide the answer to many of the legal questions connected to cross-border assistance operations, it must be borne in mind that other branches of international law might play a decisive role in shaping disaster response activities. These include the law of international organisations, especially if they have legal mandates relevant to disaster management.[84] International Human Rights Law (IHRL) is likewise absolutely central to disaster activities as it sets out rights germane to disaster relief and recovery assistance, such as the right to life, to food and water, to health, to livelihood and freedom from discrimination, that need to be respected by the affected State. Moreover, IHRL provides for limitations on the type of restrictions States can impose on individual rights to tackle the effects of disasters and other emergencies.[85]

International Humanitarian Law (IHL) must also be considered. While its application is generally restricted to situations of armed conflict, it may still play a role in facilitating disaster assistance. For example, IHL treaties allow National Red Cross and Red

[81] This is, for instance, the approach taken within the EU Civil Protection Mechanism; see Commission Implementing Decision 2014/762/EU, Art. 40.

[82] Convention on Assistance in the Case of a Nuclear Accident, Art. 10.2.b.

[83] Crawford (2012), pp. 489–491.

[84] Reference must be made to the UN and its numerous agencies, but also to regional organizations such as the EU. Article 6(f) of Treaty on the Functioning of the European Union (Lisbon Treaty), 13 December 2007, sets forth in general terms a competence of the Union 'to carry out actions to support, coordinate or supplement the actions of the member States' inter alia in the area of civil protection. The competence is then spelled out in Title XXIII of the TFEU, entitled 'civil protection' (Art. 196).

[85] Sommario (2018), pp. 99–111.

Crescent Societies, as well as national civil protection authorities, to use their respective emblems and signs during both peacetime and wartime.[86] Furthermore, when natural disasters occur alongside armed conflict, IHL will apply and its provisions on the facilitation of humanitarian assistance could also prove useful to tackle the effects of the disaster. In any case, it is useful to draw analogies with IHL when addressing issues also encountered by International Disaster Law as there are many similarities in terms of the needs of the affected population and of the possible legal obstacles that first responders might face.[87] Other relevant areas of international law include International Environmental Law (IEL), Global Health Law (with respect to State obligations concerning pandemics), and International Trade Law (especially with regard to the customs regimes), to name a few.[88]

4.4 The (Scant) International Legal Framework on Disaster Recovery

The traditional emphasis on the facilitation of assistance in the aftermath of a disaster has meant that international disaster law prioritised immediate response and initial recovery over other phases of the disaster cycle. Over the last three decades, the temporal scope of the field has expanded to encompass pre-disaster activities, particularly those focused on risk reduction and preparedness. In contrast, the post-disaster phase of recovery—and its rehabilitation and reconstruction components—has mostly been addressed by other actors and governed by separate, mainly domestic, legal, and policy frameworks.[89] However, while it is recognised that current international law does not provide a solid framework specifically dedicated to disaster recovery,[90] several legal instruments, principles, and frameworks collectively regulate and guide these activities. The regulation spans multiple branches of international law but primarily involves some disaster-specific soft law instruments and a few IEL treaties.[91]

---

[86] See, for instance, Geneva Convention for the Amelioration of the Condition of the Wounded and Sick in Armed Forces in the Field, 12 August 1949, Art. 44, and Protocol Additional to the Geneva Conventions of 12 August 1949, and Relating to the Protection of Victims of International Armed Conflicts, 8 June 1977, Art. 66.7.

[87] Venturini (2012), p. 52.

[88] For a review of the relevant areas see Fisher (2007), pp. 33–52.

[89] Pronto (2021), p. 573.

[90] Note, however, that the rehabilitation phase is explicitly mentioned in some binding disaster law agreements. For instance, Art. 17 of the ASEAN Agreement on Disaster Management and Emergency Response demands that State parties 'develop strategies and implement programmes for rehabilitation as a result of a disaster' and that they 'promote, as appropriate, bilateral, regional and international co-operation for rehabilitation as a result of a disaster'.

[91] As in other phases of the disaster cycle, IHRL maintains its relevance in informing interventions even during the recovery phase. This means that all actions must incorporate approaches that are participatory, gender-responsive, culturally sensitive, accountable, and non-discriminatory. These approaches should also adhere to the principles of universal design, accessibility, inclusiveness, and

4.4.1 Disaster-Specific Soft Law Instruments

While different non-binding documents could be considered applicable to the recovery phase, for the purpose of the present section we will focus on the contribution of the three instruments more relevant to these activities, i.e. the above-mentioned UNGA Resolution 46/182, the Sendai Framework for Disaster Risk Reduction, and the UN Sustainable Development Goals.

UNGA Resolution 46/182 addresses the coordination of United Nations humanitarian emergency assistance and is widely regarded as a landmark instrument. The resolution 'establishes the framework within which international disaster relief activities are undertaken'. Although the text does not centre specifically on recovery, it contains several provisions that are directly relevant to post-disaster rehabilitation efforts. In particular, it recognises a 'clear relationship between emergency, rehabilitation and development',[92] and emphasises that 'emergency assistance must be provided in ways that will be supportive of recovery and long-term development'.[93] Furthermore, it states that 'international cooperation and support for rehabilitation and reconstruction should continue with sustained intensity after the initial relief stage', further noting that rehabilitation presents 'an opportunity to restructure and improve facilities and services destroyed by emergencies'.[94] These provisions collectively underscore the Resolution's broader vision of integrating emergency response with longer-term recovery and development goals.

The *Sendai Framework for Disaster Risk Reduction 2015–2030* is a 15-year, voluntary, and non-binding international agreement adopted at the Third United Nations World Conference on Disaster Risk Reduction, held in March 2015 in Sendai, Miyagi, Japan.[95] It succeeds the Hyogo Framework for Action 2005–2015: Building the Resilience of Nations and Communities to Disasters.[96]The Sendai Framework has been characterised as an international disaster law instrument that simultaneously addresses disaster risk reduction (DRR), sustainable development, and climate change adaptation.[97] The text reaffirms that States bear the primary responsibility for disaster risk management, including post-disaster recovery, but also calls for shared responsibility among a broad array of

the 'do no harm' principle. See Human Rights Council, Final research-based report of the Human Rights Council Advisory Committee on best practices and main challenges in the promotion and protection of human rights in post-disaster and post-conflict situations, UN Doc A/HRC/28/76, 10 February 2015.

[92] UN Doc A/RES/46/182, para. 9.

[93] Ibid., para. 40.

[94] Ibid., para. 41.

[95] Sendai Framework for Disaster Risk Reduction 2015–2030, UN Doc A/CONF.224/CRP.1, 18 March 2015.

[96] Hyogo Framework for Action 2005–2015: Building the Resilience of Nations and Communities to Disasters, Extract from the Final Report of the World Conference on Disaster Reduction, UN Doc A/CONF.206/6, 16 March 2005.

[97] Peel and Fisher (2016), pp. 13–14.

stakeholders, including local authorities, the private sector, and civil society.[98] While it contains principles encouraging international cooperation, the primary focus is on actions to be undertaken by States within their own jurisdictions.[99]

The Sendai Framework outlines four priority areas, with Priority 4 ('Enhancing disaster preparedness for effective response and to 'build back better' in recovery, rehabilitation and reconstruction') being directly relevant to recovery efforts.[100] To implement this priority, the Framework enumerates 8 global and regional actions and 17 national- and local-level actions. Among the most significant of these for the recovery phase are

1. Adopting policies that empower public authorities to strengthen coordination and establish funding mechanisms for recovery and reconstruction.[101]
2. Promoting both horizontal and vertical coordination across institutions and levels of governance.[102]
3. Integrating disaster risk reduction and sustainable development into recovery efforts, including in land use planning, infrastructure relocation, and the application of improved structural standards.[103]
4. Facilitating the sharing of expertise and experiences in recovery programming, building upon lessons from the Hyogo Framework.[104]
5. Enhancing recovery frameworks through the inclusion of psychosocial support and mental health services.[105]

Through these provisions, the Sendai Framework establishes a comprehensive and forward-looking approach to post-disaster recovery, positioning it as a key component of broader resilience-building and development agendas.

The United Nations Sustainable Development Goals 2015 ('SDGs') were adopted in a resolution by the UNGA as part of the 2030 Agenda for Sustainable Development.[106] These goals are non-binding and are only generally applicable to DRM. However, given their prominence and near-universal acceptance, they are included here as a key international soft law instrument. Some of the SDGs are particularly relevant to disaster recovery. Goal 9 asks States, inter alia, to build 'resilient infrastructure', while Goal 11 demands that

[98] Sendai Framework, para. 19(e).

[99] Ibid., para. 27.

[100] Ibid., para. 33.

[101] Ibid., para. 33.e.

[102] Ibid., para. 33.i.

[103] Ibid., para. 33.j.

[104] Ibid., paras. 33.j–k.

[105] Ibid., para. 33.o.

[106] UNGA, Transforming our world: the 2030 Agenda for Sustainable Development, UN Doc A/RES/70/1, 21 October 2015.

they '[m]ake cities and human settlements inclusive, safe, resilient and sustainable'.[107] Lastly, Goal 13 requires that they take 'urgent action to combat climate change and its impacts'.[108] These three SDGs taken together underline the need to bring reconstruction and rehabilitation efforts in line with the development agenda.

4.4.2 International Environmental Law and Disaster Recovery

While there are no IEL instruments that deal exclusively with the post-disaster recovery phase, several instruments are highly relevant due to their emphasis on environmental protection, climate change, and sustainability, all of which are integral to responsible and resilient recovery practices. Among the key instruments in this respect are the 1992 Rio Declaration on Environment and Development,[109] the United Nations Framework Convention on Climate Change (UNFCCC)[110] and, related to the latter, the 2015 Paris Agreement.[111]

The Rio Declaration, alongside its accompanying action plan, the Agenda 21,[112] represents the core framework for international environmental governance. Although considered a soft law instrument, it has had a profound influence on the development of international treaties and on both domestic and international environmental policies.[113] The Declaration sets out 27 principles of sustainable development. Some are particularly relevant to disaster recovery and the concept of 'build back better'. In particular, Principle 15 articulates the precautionary approach, providing that where there is a risk of serious or irreversible environmental harm, a lack of full scientific certainty shall not be used as a reason to postpone cost-effective measures to prevent environmental degradation.[114] Also of relevance is Principle 17, which calls for environmental impact assessments to be conducted for activities that are likely to have a significant adverse effect on the environment and are subject to the decision of a competent national authority.[115] Together, these principles underscore the importance of integrating environmental considerations

[107] In particular, Target 11.b encourages recovery planning as part of holistic disaster risk management, in line with the Sendai Framework, which includes recovery as a core pillar.

[108] Target 13.1 calls upon States to strengthen resilience and adaptive capacity to climate-related hazards and natural disasters and invites them to implement adaptive recovery efforts after disasters and climate-related shocks.

[109] Rio Declaration on Environment and Development, UN Doc A/CONF.151/26 (Vol. I), 12 August 1992.

[110] United Nations Framework Convention on Climate Change, 4 June 1992.

[111] UNFCCC Conference of the Parties, Adoption of the Paris Agreement, UN Doc FCCC/CP/2015/L.9/Rev./1, 12 December 2015.

[112] Agenda 21, UN Doc A/CONF.151/26 (Vol. II), 13 August 1992.

[113] Morin et al. (2023), pp. 486–487.

[114] Rio Declaration, Principle 15, p. 6.

[115] Ibid.

into recovery strategies, not only to avoid further harm but also to enhance resilience and sustainability in the aftermath of disasters.

The UNFCCC is the foundation of the UN climate change regime and aims at the 'stabilisation of greenhouse gas concentrations in the atmosphere at a level that would prevent dangerous anthropogenic interference with the climate system'.[116] It establishes a number of institutions to facilitate its work, including its Conference of the Parties (COP), mandated to 'make the decisions necessary to promote the effective implementation of the Convention'.[117] In 2015, the State parties to the UNFCCC adopted the Paris Agreement, whose core aim is to tackle climate change by '[h]olding the increase in the global average temperature to well below 2°C above pre-industrial levels and pursuing efforts to limit the temperature increase to 1.5°C above pre-industrial levels'.[118]

Of critical relevance to recovery is the parties' commitment under Article 7 of the Paris Agreement to 'establish the global goal on adaptation of enhancing adaptive capacity, strengthening resilience and reducing vulnerability to climate change'. The wording appears to suggest that reconstruction efforts need to be aligned with the requirements of climate change adaptation. Another relevant feature of the Paris Agreement is its 'loss and damage' mechanism. Article 8 of the treaty acknowledges the need to avert, minimise, and address the 'loss and damage associated with the impacts of climate change, including extreme weather events and slow-onset events'. These impacts often cause irreversible harm and long-term disruption, which are core concerns of disaster recovery processes. The same provision establishes the Warsaw International Mechanism for Loss and Damage, which focuses on enhancing the understanding of climate impacts, strengthening dialogue and coordination, and promoting action and support, including finance, technology, and capacity-building. These functions align closely with disaster recovery needs, especially where countries lack the institutional capacity to respond and rebuild effectively. At COP27 in 2022, the State parties agreed to establish a Loss and Damage Fund, with the goal of providing financial assistance to developing countries facing climate-induced losses.[119] Although the details are still being operationalised, the fund is expected to support post-disaster recovery, including reconstruction of infrastructure, restoration of livelihoods, and rehabilitation of affected ecosystems.

---

[116] UNFCCC, Art. 2.

[117] UNFCCC, Art. 7.

[118] Paris Agreement, Art. 2.1.a.

[119] UNFCCC, Decision 2/CP.27: Funding arrangements for responding to loss and damage associated with the adverse effects of climate change, including a focus on addressing loss and damage (2022) UN Doc FCCC/CP/2022/10/Add.1, 20 November 2022, and UNFCCC, Decision 2/CMA.4: Funding arrangements for responding to loss and damage associated with the adverse effects of climate change, including a focus on addressing loss and damage (2022) UN Doc FCCC/PA/CMA/2022/10/Add.1, 20 November 2022.

4.5 Conclusion: A New Comprehensive Treaty on Disaster Risk Management?

As evidenced while describing the international legal framework pertaining to DRM, the international community still lacks a comprehensive, legally binding instrument of universal scope setting out rules applicable in the various phases of the disaster cycle. However, recent events, including the COVID-19 pandemic, have awakened an appetite for such an instrument, which might soon become a reality. The initial driver in this process was the 2004 Indian Ocean tsunami, a disaster which was transboundary in nature and resulted in a massive loss of life, exposing the international community's lack of preparedness for calamities of such scale. In the aftermath of the event, the ILC initiated its examination of the topic 'Protection of Persons in the Event of Disasters', formally including it in its programme of work in 2007 and appointing Eduardo Valencia-Ospina as Special Rapporteur. As is known, in 2016 the ILC adopted a set of DAs[120] and the document was then transmitted to the UNGA, together with a recommendation that an international convention be concluded on the basis of its text.[121] The recommendation of the Commission has since then been considered several times in the Sixth (Legal) Committee of the UNGA. In December 2024, the UNGA adopted by consensus Resolution 79/128, through which it decided to 'elaborate and conclude a legally binding instrument on the protection of persons in the event of disasters (...) by the end of 2027 at the latest'.[122] The UNGA also invited Governments to submit to the UN Secretary-General any proposal for amendments to the DAs by the end of 2025, 'with a view to preparing the consolidated text that will serve as the basis for the negotiations of the legally binding instrument'.[123]

In their current version, the DAs are made up of 18 provisions, preceded by a preamble. The preamble aims at providing a conceptual framework, setting out the general context in which the topic has been elaborated and furnishing the rationale for the text. DA 1 and DA 2 set out the scope and purpose of the text, which is 'to facilitate the adequate and effective response to disasters and reduction of the risk of disasters, so as to meet the essential needs of the persons concerned, with full respect for their rights'. DA 3 includes definitions for the various terms used in the document, including a definition of the notion of 'disaster'.[124] DAs 4 to 6 address the relationship between the victims of disasters and assisting actors, linking relief action to respect for human rights and humanitarian principles. On the other hand, DAs 7 to 17 focus on the relationship between the affected State and assisting actors, addressing key issues such as the duties and role of the affected State in disaster response and disaster risk reduction activities, consent to

120 Yearbook of the International Law Commission, 2016, vol. II (Part Two) para. 48.

121 Ibid., para. 47.

122 UNGA, Protection of persons in the event of disasters, UN Doc A/RES/79/128, 12 December 2024, OP 4.

123 Ibid., OP 6.

124 See Sect. 4.2.1.

assistance, offers and requests of assistance, and operational provisions governing relief operations. Finally, its final provision (DA 18) articulates the relationship between the DAs and other rules of international law, including IHL.

While we do have a roadmap which should lead to the conclusion of an agreement by the end of 2027, negotiations will not be easy as a series of legal issues divide States and striking a compromise might prove arduous.[125] One problematic provision is DA 11, which imposes on the affected State a duty to seek assistance from external actors if the disaster 'manifestly exceeds its national response capacity'. China, for instance, argued that such an obligation is not part of customary international law, suggesting that it might not favour the provision.[126] Also the object of some criticism is DA 13, whose second paragraph established that '[c]onsent to external assistance shall not be withheld arbitrarily'. A number of States have expressed perplexity as they maintain that the term 'arbitrarily' is not sufficiently clear.[127]

Despite these and other controversies, the time is ripe for the adoption of a treaty that could provide a degree of certainty about what governments can expect from other governments and humanitarian organisations following a disaster and clarify what facilitation measures must be adopted if external assistance is provided. It would also bring some order to the regulation of disaster management, with existing provisions being scattered across many different instruments of varying status. The duty is unavoidable as disasters will only increase as a consequence of climate change, as will the number of victims and the material damage caused. An effective legal framework is an essential point of departure to reduce the risk of disasters and, when they occur, to provide quick and effective relief to affected communities.

Key Sentences

- **The Central Role of State Sovereignty**—International disaster response is grounded in the sovereignty of the affected State, meaning that external assistance is only permissible with that State's consent, which retains primary responsibility for disaster management.

[125] See Bartolini (2024).

[126] UNGA, Sixth Committee. Summary record of the 20th meeting, UN Doc A/C.6/71/SR.20, 11 November 2016, para. 64.

[127] See, for instance, the positions of Israel, UNGA, Sixth Committee. Summary record of the 23rd meeting, UN Doc A/C.6/66/SR.23, 14 November 2011, para. 33; and of the United Kingdom, ibid. para. 45.

- **The Lack of a Universally Accepted Definitions**—There is no universally accepted definition of 'disaster', and various legal instruments use differing criteria, such as the magnitude of damage, the nature of the threat, and the coping capacity of the State, to trigger legal obligations.
- **The Scarcity of Binding Instruments Concerning Disaster Recovery**—While disaster response law is relatively well developed, disaster recovery remains under-regulated, relying mostly on non-binding frameworks like the Sendai Framework and the Sustainable Development Goals.
- **The Fragmentation of the Legal Framework**—A multitude of legal issues (including entry of personnel, visas, customs, liability, recognition of qualifications) arise in cross-border disaster response and are addressed by a patchwork of treaties and soft law instruments.
- **The Future Adoption of a Flagship Treaty**—Efforts are underway to develop a comprehensive and binding international treaty on disaster risk management, following the UNGA's 2024 resolution to negotiate such an instrument based on the International Law Commission's *Draft articles on the protection of persons in the event of disasters*.
- **The Call for a More Effective Legal Framework**—As climate change increases the frequency and severity of disasters, the need for a unified, predictable, and legally robust global framework is more urgent than ever.

References

Adinolfi G (2012) Customs obstacles to relief consignments under international disaster response law. In: de Guttry A et al (eds) International disaster response law. Springer, pp 533–550

Bartolini G (2018) A taxonomy of disasters in international law. In: Zorzi Giustiniani F et al (eds) Routledge handbook of human rights and disasters. Routledge, pp 10–26

Bartolini G (2024) Toward a universal treaty on 'protection of persons in the event of disasters'. EJIL Talk. https://www.ejiltalk.org/toward-a-universal-treaty-on-protection-of-persons-in-the-event-of-disasters/

Crawford JR (2012) Brownlie's principles of public international law, 8th edn. Oxford University Press

de Guttry A (2012) Surveying the law. In: de Guttry A et al (eds) International disaster response law. Springer, pp 3–44

Fisher D (2007) Law and legal issues in international disaster response: a desk study. International Federation of Red Cross and Red Crescent Societies

Focarelli C (2021) Duty to protect in cases of natural disasters. In: Peters A (ed) Max Planck encyclopedia of public international law. Oxford University Press

Gestri M (2012) EU disaster response law: principles and instruments. In: de Guttry A et al (eds) International disaster response law. Springer, pp 105–128

Hartman C, Squires GD (eds) (2006) There is no such thing as a natural disaster: race, class, and hurricane Katrina. Routledge

Morin JF et al (2023) The survival of the weakest: the echo of the Rio summit principles in environmental treaties. Environ Polit 33(3):486–507

Peel J, Fisher D (2016) International law at the intersection of environmental protection and disaster risk reduction. In: Peel J, Fisher D (eds) The role of international environmental law in disaster risk reduction, pp 1–25

Pronto AN (2021) International disaster law. In: Geiss R, Melzer N (eds) The Oxford handbook of the international law of global security. Oxford University Press, pp 566–584

Shaw MN (2021) International law. Chapter 12, immunities from jurisdiction. Cambridge University Press, pp 601–676

Silingardi S (2012a) The status of emergency workers. In: Guttry A de et al (eds) International disaster response law. Springer, pp 551–584

Silingardi S (2012b) From theory to practice: the role of disaster response missions. In: Guttry A de et al (eds) International disaster response law. Springer, pp 465–484

Sommario E (2018) Limitation and derogation provisions in international human rights law treaties and their use in disaster settings. In: Zorzi Giustiniani F et al (eds) Routledge handbook of human rights and disasters. Routledge, pp 98–118

Venturini G (2012) International disaster response law in relation to other branches of international law. In: de Guttry A et al (eds) International disaster response law. Springer, pp 45–64

Zorzi Giustiniani F (2021) International law in disaster scenarios. Applicable Rules and Principles. Springer

Open Access This chapter is licensed under the terms of the Creative Commons Attribution 4.0 International License (http://creativecommons.org/licenses/by/4.0/), which permits use, sharing, adaptation, distribution and reproduction in any medium or format, as long as you give appropriate credit to the original author(s) and the source, provide a link to the Creative Commons license and indicate if changes were made.

The images or other third party material in this chapter are included in the chapter's Creative Commons license, unless indicated otherwise in a credit line to the material. If material is not included in the chapter's Creative Commons license and your intended use is not permitted by statutory regulation or exceeds the permitted use, you will need to obtain permission directly from the copyright holder.

The International Legal Framework for Disaster Prevention and Preparedness Activities

5

Riccardo Luporini

Abstract

This chapter surveys a range of legal instruments to outline the scope and content of international obligations and standards governing disaster prevention and preparedness. It recognises that the measures required of public authorities and other relevant actors are numerous and vary depending on the nature of the hazard and the specific circumstances. Nonetheless, the chapter highlights a set of core activities that have gained prominence in international legal and institutional practice. These include the adoption of dedicated regulatory frameworks, the conduct of risk assessments, the development of contingency plans, the sharing of information, the delivery of training and awareness-raising programmes, and the notification of hazardous events. Despite the fragmented character of the existing international legal framework, prevention and preparedness are gaining growing recognition in multilateral treaties, soft law, and regional practice. These developments offer grounds for measured optimism regarding the potential of international law and institutions to contribute meaningfully to a more disaster-resilient future.

Keywords

International disaster law • Prevention • Preparedness • Disaster risk reduction • Early warning systems

R. Luporini (✉)
Scuola Superiore Sant'Anna, Pisa, Italy
e-mail: riccardo1.luporini@santannapisa.it

© The Author(s) 2026

E. Sommario and P. Billing (eds.), *International Legal and Institutional Frameworks for Disaster Risk Management*, Civil Protection, Disaster Management, and Resilience 4,
https://doi.org/10.1007/978-3-032-11806-6_5

5.1 Introduction

Disaster prevention has been historically undervalued compared to disaster response, not least from an international regulatory perspective. In its efforts to establish universal legal instruments and common standards for disaster management, the international community has initially focused on disaster response, particularly facilitating the delivery of international assistance following major events that overwhelm national capacities.[1] However, over time, the prevention of and preparedness for disasters gained increasing recognition.[2] Today, obligations in these areas are reflected in several binding multilateral sectoral treaties, as well as in various regional arrangements, notably in the European Union (EU), but also in other regions. The Sendai Framework for Disaster Risk Reduction 2015–2030 (Sendai Framework),[3] while formally a soft law instrument and thus not legally binding, has nonetheless become the leading global framework for disaster prevention and preparedness across sectors. It offers a comprehensive set of standards and priorities for action that guide the conduct of States and other relevant actors.

Against this background, the present chapter aims to introduce readers to the multifaceted and rather fragmented international legal framework regulating disaster prevention and preparedness activities.

The introduction to this volume has already outlined the terminology used throughout. To briefly recall, for the purposes of this chapter, *prevention* refers to all activities designed to proactively reduce the risk of disasters, while *preparedness* encompasses actions aimed at building the institutional, community, and individual capacities necessary

[1] Reference can be made to the Convention Establishing an International Relief Union, 12 July 1927. Although the Convention entered into force in 1932 with 29 State parties, it ultimately failed to meet its objectives, largely due to the broader institutional shortcomings of the League of Nations. Decades later, in 1971, the United Nations (UN) took steps to strengthen international disaster response by appointing a Disaster Relief Coordinator and establishing the UN Disaster Relief Organization (UNDRO), see UN General Assembly (UNGA), Res 2816 (XXVI) 'Assistance in cases of natural disaster and other disaster situations' (6 December 1971) UN Doc A/RES/2816. In 1984, the UN Economic and Social Council, acting on the initiative of the Disaster Relief Coordinator, undertook efforts to draft a convention aimed at expediting the delivery of emergency assistance. The initiative was eventually abandoned.

[2] A shift in focus began to take shape in 1987 with the proclamation of the International Decade for Natural Disaster Reduction (1990–1999), see UNGA, Res 42/169 'International Decade for Disaster Reduction' (11 December 1987) UN Doc A/RES/42/169. This initiative marked a growing recognition of the need for a more proactive and coordinated approach to disaster risk reduction. In 2000, the Decade was succeeded by the International Strategy for Disaster Reduction, established through UNGA, Res 54/219 'International Decade for Natural Disaster Reduction: successor arrangements' (3 February 2000) UN Doc A/RES/54/219.

[3] UNGA, Res 69/283 'Sendai Framework for Disaster Risk Reduction 2015-2030' (23 June 2015) UN Doc RES/69/283.

to manage disasters effectively.[4] Importantly, this terminology resonates with that adopted at the EU level. According to Decision 1313/2013/EU on the Union Civil Protection Mechanism (UCPM), prevention means 'any action aimed at reducing risk or mitigating adverse consequences of a disaster for people, the environment and property, including cultural heritage', while preparedness is defined as 'a state of readiness and capability of human and material means, structures, communities and organisations enabling them to ensure an effective rapid response to a disaster, obtained as a result of action taken in advance'.[5] Prevention and preparedness—often along with 'mitigation'—are typically grouped under the broader concept of *disaster risk reduction* (DRR). DRR encompasses 'all activities aimed at preventing new and reducing existing disaster risk, and managing residual risk' and will serve as an additional key reference concept throughout this chapter.[6]

Some criticisms of the conceptual model of the 'disaster cycle' have already been noted in the introduction to this volume.[7] Especially relevant for our purposes, the boundaries between disaster phases are increasingly blurred, particularly in the case of slow-onset events, such as those associated with climate change, that unfold gradually and do not fit neatly into distinct temporal stages.[8] Nevertheless, for reasons of clarity and consistency, this chapter adopts the traditional disaster cycle approach, focusing on the so-called 'pre-disaster phase' and distinguishing between the international legal norms and standards regulating disaster prevention on the one hand, and preparedness on the other. This distinction is not always straightforward, as certain measures may fall under both prevention and preparedness, or it may be unclear to which category they belong. A clear example is the implementation of effective *early warning systems*,[9] which typically encompass

[4] See the definitions adopted by the Open-Ended Intergovernmental Expert Working Group on Indicators and Terminology Relating to Disaster Risk Reduction, as endorsed by the UNGA: UNGA, Res 71/644 'Report of the Open-Ended Intergovernmental Expert Working Group on Indicators and Terminology Relating to Disaster Risk Reduction' (1 December 2016) UN Doc RES/71/644. The same glossary is available online at https://www.undrr.org/drr-glossary/terminology.

[5] Decision 1313/2013/EU of the European Parliament and of the Council on a Union Civil Protection Mechanism, OJL 347 (17 December 2013) (UCPM Decision), Article 4,4 and 4,3, respectively.

[6] Similarly, 'disaster risk management' is defined as 'the application of disaster risk reduction policies and strategies to prevent new disaster risk, reduce existing disaster risk and manage residual risk, contributing to the strengthening of resilience and reduction of disaster losses'. 'Disaster risk governance' is defined as 'the system of institutions, mechanisms, policy and legal frameworks and other arrangements to guide, coordinate and oversee disaster risk reduction and related areas of policy'. See UNGA, Report of the Open-Ended Intergovernmental Expert Working Group on Indicators and Terminology Relating to Disaster Risk Reduction. See also Zorzi Giustiniani (2018).

[7] Some scholars and practitioners prefer to use the concept of the 'disaster risk reduction continuum', see, for instance, Spiekermann et al. (2015).

[8] See, among others, Staupe-Delgado and Rubin (2022).

[9] An early warning system is defined as 'an integrated system of hazard monitoring, forecasting and prediction, disaster risk assessment, communication and preparedness activities systems and processes that enables individuals, communities, governments, businesses and others to take timely

elements of both. Components such as risk assessments, disseminating disaster risk information and warnings, and developing contingency plans will therefore be separated and analysed individually.

Accordingly, the chapter is structured as follows. Section 5.2 illustrates the content and scope of the international obligation to prevent disasters, briefly recalling relevant international legal norms and identifying key prevention measures. Section 5.3 applies the same approach to disaster preparedness. Section 5.4 concludes the chapter and highlights some recent developments.

5.2 The International Obligation to Prevent Disasters

As noted in Sommario's chapter in the present volume, unlike other more consolidated areas of international law, disaster risk management has not yet been codified in a comprehensive, universal treaty. After the failure of certain early attempts,[10] the idea of elaborating a universal treaty to regulate disaster prevention, response, and recovery was considered impractical and was indefinitely postponed. As a result, international disaster law has developed as a 'rather scattered and heterogeneous collection of instruments',[11] either sectoral or of a regional or sub-regional character. This means, in the first place, that at present there is no binding universal instrument enshrining a duty to prevent disasters. While some recent relevant developments in this respect will be discussed in the conclusions of the chapter, it is important now to stress that, despite the lack of a universal binding written norm enshrining the duty to prevent disasters, such an obligation can be derived from a very diverse but still solid set of international instruments that deal, from different angles, with disaster prevention.

This resonates in the first place with the *principle of prevention*, which has long been recognised across various branches of international law, particularly international environmental law, and requires States to take appropriate measures to avert significant harm.[12] This has been affirmed by the International Court of Justice (ICJ)—principal judicial body of the UN—in key decisions such as *Pulp Mills on the River Uruguay* and *Gabčíkovo-Nagymaros*, which confirmed that States must anticipate, prevent, and mitigate environmental harm, especially when it may have transboundary consequences.[13] Closely

action to reduce disaster risks in advance of hazardous events', see UNGA, Report of the Open-Ended Intergovernmental Expert Working Group on Indicators and Terminology Relating to Disaster Risk Reduction.

[10] See the Convention Establishing an International Relief Union and the draft convention on expediting the delivery of emergency assistance.

[11] Fisher (2007), p. 353.

[12] See Duvic-Paoli (2018), Aronsson-Storrier (2019) and Cubie (2024).

[13] See ICJ *Pulp Mills on the River Uruguay (Argentina v Uruguay)*, Judgement of 20 April 2010 and ICJ, *Gabčíkovo-Nagymaros Project (Hungary v Slovakia)*, Judgement of 5 February 1997.

related is the *precautionary principle*, often invoked in environmental law, which requires States to act to prevent harm to human health or the environment even in the face of scientific uncertainty.[14]

Alongside these principles, various *multilateral environmental agreements* contribute significantly to establishing a duty to prevent disasters. These treaties primarily address environmental degradation processes, which exacerbate natural hazards over time. For instance, the UN Convention to Combat Desertification (UNCCD)[15] provides a legal framework for addressing land degradation, desertification, and drought, particularly in vulnerable regions such as sub-Saharan Africa, where these phenomena can meet the severity threshold of disasters. Likewise, the Ramsar Convention on Wetlands[16] and the Convention on Biological Diversity (CBD)[17] promote ecosystem resilience in the face of climate change and biodiversity loss. Instruments such as the Convention on International Trade in Endangered Species of Wild Fauna and Flora (CITES)[18] and the Convention on the Conservation of Migratory Species of Wild Animals (CMS)[19] also contribute indirectly to reducing risks, for example, by addressing zoonotic spillover potential.

A particularly important area for disaster prevention is *international climate change law*, given the increasing frequency and intensity of climate-related hazards. The international climate change regime comprises three core treaties: the UN Framework Convention on Climate Change (UNFCCC),[20] the Kyoto Protocol,[21] and the Paris Agreement.[22] Since the adoption of the UNFCCC, a complex institutional architecture has evolved, comprising the UNFCCC Secretariat, the Conference of the Parties (COP), and a range of subsidiary organs. Under this regime, States are required to undertake both climate change mitigation measures—namely, the reduction of greenhouse gas emissions—and adaptation measures.[23] Both sets of actions intersect with disaster prevention. Mitigation contributes

---

[14] See Schröder (2014) and Sadeleer (2016).

[15] United Nations Convention to Combat Desertification in those Countries Experiencing Serious Drought and/or Desertification, Particularly in Africa, 14 October 1994.

[16] Convention on wetlands of international importance especially as waterfowl habitat, 2 February 1971.

[17] Convention on Biological Diversity, 5 June 1992.

[18] Convention on International Trade in Endangered Species of Wild Fauna and Flora, 3 March 1973.

[19] Convention on the conservation of migratory species of wild animals, 23 June 1979.

[20] United Nations Framework Convention on Climate Change, 9 May 1992.

[21] Kyoto Protocol to the United Nations Framework Convention on Climate Change, 11 December 1997.

[22] Paris Agreement, 12 December 2015.

[23] The Intergovernmental Panel on Climate Change (IPCC) provides the following definition of adaptation: 'In human systems the process of adjustment to actual or expected climate and its effects, in order to moderate harm or exploit beneficial opportunities. In natural systems, the process of adjustment to actual climate and its effects; human intervention may facilitate adjustment to expected climate'. See IPCC (2023), p. 120.

to reducing the hazard component of the risk equation by lessening the long-term frequency and intensity of climate-related impacts. Adaptation, on the other hand, targets the remaining components—exposure, vulnerability, and coping capacity—and significantly overlaps with DRR efforts.[24]

Climate-related hazards are not the only major concern for future disaster prevention efforts. The experiences of Ebola and COVID-19 have brought to the forefront the need for stronger international legal frameworks for *epidemic and pandemic risk*. The main legal instrument regulating public health emergencies is the International Health Regulations (IHR), adopted in 2005 under the auspices of the World Health Organization (WHO) and revised in 2024.[25] The IHR set out rights and obligations for States in preventing, detecting, and responding to health events with cross-border implications. They define criteria for declaring a 'public health emergency of international concern', require States to notify of such events, and mandate the establishment of national focal points. This regulatory framework has been recently strengthened with the adoption of the Pandemic Agreement. Approved by the World Health Assembly, the treaty—not yet in force—governs prevention, preparedness, and response activities in relation to pandemic emergencies and is grounded in cooperation, equity, and the 'One Health' approach.[26]

Another distinct and relevant area to highlight is *international human rights law*. The link between disasters and the protection of human rights is today well established in its factual, legal, and operational dimensions.[27] International and regional human rights bodies have progressively addressed disaster-related issues. Among others, the European Court of Human Rights (ECtHR) has delivered a few landmark judgements affirming that States may incur responsibility under the European Convention on Human Rights (ECHR)[28] for failing to adopt adequate risk reduction measures in relation to both human-made and natural hazards.[29] This growing body of jurisprudence underscores the principle that public authorities have a legal obligation to take appropriate preventive

---

[24] See Natoli (2019) and Luporini (2025). More recently, a third pillar of climate action has emerged: 'addressing loss and damage associated with the adverse impacts of climate change'. This area, too, intersects closely with disaster management, particularly in relation to post-disaster response, recovery, and support for affected communities. See Bhandari et al. (2025).

[25] International Health Regulations, 23 May 2005. See also Habibi et al. (2025) and Toscano-Rivalta (2020).

[26] World Health Assembly, '*WHO Pandemic Agreement*' (20 May 2025) WHA78.1. Before the treaty is opened for signature, an Intergovernmental Working Group will have to negotiate and adopt an annex concerning the Pathogen Access and Benefit-Sharing (PABS) system, governed by Article 12 of the treaty. The latter will enter into force upon reaching 60 ratifications.

[27] See Kälin (2012), Zorzi Giustiani et al. (eds) (2018) and Sommario and Venier (2018).

[28] European Convention on Human Rights, 4 November 1950.

[29] ECtHR, *Lopez Ostra v Spain*, Judgement of 9 December 1994; *Guerra and others v Italy*, Judgement of February 1998; *Öneryildiz v Turkey*, Judgement of 30 November 2004; *Budayeva v Russia*, Judgement of 20 March 2008; *Kolyadenko v Russia*, Judgement of 9 July 2012.

and preparedness measures to protect human rights against the serious threats posed by disasters.

The resulting obligation to prevent disasters is best characterised as one of conduct or '*due diligence*': States must undertake all necessary and reasonable measures to reduce disaster risk, acknowledging that disasters cannot be entirely prevented.[30] A certain level of residual risk must be accepted, and measures should aim to respond to such risk and recover when disasters do occur.[31] States enjoy discretion in how to regulate disaster prevention in their jurisdictions. However, this discretion is constrained by certain parameters derived from various international instruments. Among these are the principles of non-retrogression and non-discrimination. The former requires that measures should consistently aim to improve protection levels and avoid any deterioration in the enjoyment of rights[32]; the latter ensures that such measures apply equally to all, with particular attention to those in vulnerable or disadvantaged situations.[33] In some contexts, especially concerning certain human-made hazards, international treaties have indirectly extended the duty of prevention to private operators, such as license holders of nuclear installations or masters of maritime vessels.[34]

The activities required of public authorities and other relevant actors to meet their prevention obligations are numerous and closely tied to the nature of the specific hazard and the circumstances at hand. The following paragraphs outline a selection of four key preventive measures: (i) establishing national and local legal and institutional frameworks for disaster prevention; (ii) conducting disaster risk assessments; (iii) ensuring access to information and public participation; and (iv) facilitating cross-border information exchange with other States and international bodies.

5.2.1 National and Local Legal and Institutional Frameworks for Disaster Prevention

The adoption and implementation of national and local legislative and administrative frameworks for disaster prevention represent the first and most essential component of the international obligation to prevent disasters. A well-functioning regulatory and institutional framework is a necessary precondition for the effective adoption of preventive measures. While international law is often seen primarily as a means of fostering inter-State cooperation, it also plays a crucial role in establishing minimum standards that States are expected to meet within their own jurisdictions. Moreover, although these obligations

[30] See Ollino (2022).

[31] See Sommario's chapter in the present volume for an examination of such measures.

[32] See Warwick (2019), Mitchell and Munro (2023).

[33] Casolari (2018).

[34] See Convention on Nuclear Safety, 26 October 1996, Art 9 and International Convention on Oil Pollution Preparedness, Response and Co-operation, 30 November 1990, Art 4.

are primarily fulfilled at the domestic level, they also perform an inter-State function by contributing to the reduction of transboundary disaster risks.

The Sendai Framework, under Priority 2 on 'Strengthening disaster risk governance to manage disaster risk', calls upon States to adopt national and local laws, regulations and public policies for disaster prevention. Such regulatory instruments must meet certain qualitative standards. They should provide for the planning of appropriate measures—whether structural (e.g. seawalls, flood barriers), non-structural (e.g. training programmes, public awareness campaigns), or combined (e.g. early warning systems). They should be supported by sufficient financial resources to guarantee the effective implementation of these measures.[35] A comprehensive regulatory framework also presupposes the establishment of an appropriate institutional architecture, with one or more bodies mandated to implement, coordinate, and monitor its application. The regulatory framework and the measures to be taken must be continuously reviewed and updated to ensure ongoing improvement and responsiveness to emerging risks. Moreover, the framework should promote integration and synergies with instruments governing other key policy sectors—such as urban planning, water management, public health, and climate change adaptation—all of which are crucial to effective disaster prevention.[36] Notably, various reports indicate that the number of national laws and policy frameworks addressing disaster prevention, or more broadly, DRR, has increased in recent years.[37]

Public authorities are also expected to adopt specific regulatory frameworks addressing particular types of disaster risk. This obligation is reflected in several international conventions that apply to specific sectors. In the case of human-made hazards, such obligations arise under treaties concerning the prevention of industrial accidents, nuclear safety, and the management of other hazardous activities. For example, the Convention on Nuclear Safety requires each State Party to establish a legislative and regulatory framework to govern the safety of nuclear installations. This framework must include a licensing system, provisions for regulatory inspections and assessments, and the designation of a competent regulatory authority.[38]

As previously mentioned, certain conventions also impose indirect obligations on non-State actors—typically private entities involved in managing hazardous installations. In such cases, domestic legal frameworks should incorporate adequate liability regimes to ensure that private actors can be held accountable for failing to meet their obligations under the relevant regulatory system.[39]

Similar regulatory efforts are also increasingly required in relation to disaster risks arising from natural hazards. This is affirmed in some international instruments. For

[35] Sendai Framework, para 27.

[36] Ibid.

[37] See UNDRR (2025).

[38] Convention on Nuclear Safety, Arts 4, 7, and 8.

[39] Ibid, Art 9. See also International Convention on Oil Pollution Preparedness, Response and Co-operation, Art 4.

instance, the Guidelines for the Domestic Facilitation and Regulation of International Disaster Relief and Initial Recovery Assistance (IDRL Guidelines), developed by the International Federation of Red Cross and Red Crescent Societies (IFRC), emphasise in the section on 'Early Warning and Preparedness' that 'States should adopt comprehensive legal, policy and institutional frameworks and planning for disaster prevention, mitigation, preparedness, relief and recovery'.[40]

The obligation to establish a functioning regulatory framework to protect individuals from disaster-related risks has also been affirmed by international human rights courts and bodies. Notably, the ECtHR has emphasised that a functioning regulatory and institutional framework is crucial, as it provides the foundation for identifying, adopting, and monitoring the specific actions necessary to protect individuals against foreseeable risks.[41]

5.2.2 Disaster Risk Assessments

Once a functioning regulatory and institutional framework is in place, the primary measure for effective prevention is the conduct of disaster risk assessments. Disaster risk assessment is defined as 'a qualitative or quantitative approach to determine the nature and extent of disaster risk by analysing potential hazards and evaluating existing conditions of exposure and vulnerability that together could harm people, property, services, livelihoods, and the environment on which they depend'.[42] This process involves identifying hazards and analysing their characteristics, such as location, intensity, frequency, and probability, alongside assessing the levels of exposure and vulnerability in their various dimensions, and evaluating coping capacities under different risk scenarios.

Under international law, the duty to prevent transboundary harm encompasses the obligation to conduct risk assessments, [43] as well as to undertake environmental impact assessments in relation to proposed projects or activities.[44]

Several binding sectoral instruments also mandate risk assessments. For example, the Convention on the Prevention of Major Industrial Accidents and the Chemicals Convention, both adopted under the auspices of the International Labour Organization (ILO), require industrial hazard assessments, which are often the responsibility of private operators or employers.[45] Similarly, the IHR requires States to build capacities to detect,

[40] See IFRC (2017).

[41] See, among others, ECtHR, *Budayeva v Russia*, para 129.

[42] UNGA, Res 71/644.

[43] UNGA, Draft Articles on Prevention of Transboundary Harm from Hazardous Activities. Text adopted by the International Law Commission at its fifty-third session in 2001, UN Doc A/56/10.

[44] Epiney (2009).

[45] Prevention of Major Industrial Accidents Convention (ILO Convention 174), 22 June 1993, Art 9. Chemicals Convention (ILO Convention 170), 25 June 1990.

assess, and report public health events, emphasising risk assessment as a key tool in managing public health emergencies of international concern. In the climate context, Article 7 of the Paris Agreement identifies climate risk and vulnerability assessments as essential components of national adaptation planning and action.

Regionally, particularly in the EU, instruments such as the Floods Directive set out specific obligations for Member States, including a three-step process: preliminary flood risk assessment, flood hazard and risk mapping, and the development of flood risk management plans. These are designed to enhance prevention, protection, and preparedness.[46]

As for the features of disaster risk assessments, international instruments often provide States and, where applicable, private actors with considerable discretion. Nevertheless, emerging guidance, particularly from the Sendai Framework, which places 'understanding disaster risk' as its first priority for action,[47] promotes a holistic approach. This means that risk assessments should not only address the hazard itself, but also consider other critical dimensions of risk, especially vulnerabilities affecting individuals, communities, and ecosystems.

Finally, international cooperation plays an essential role in enabling effective disaster risk assessments, particularly for hazards with transboundary or global implications, such as pandemics or climate-induced disasters. A prominent initiative in this regard is the 'Early Warnings for All' programme, launched within the UN system. Co-led by the World Meteorological Organization (WMO), the United Nations Office for Disaster Risk Reduction (UNDRR), the International Telecommunication Union (ITU), and the IFRC, the initiative aims to ensure that everyone is protected by multi-hazard early warning systems by the end of 2027. It highlights the vital role of coordinated assessments and shared data in building global resilience to complex and evolving risks.[48]

5.2.3 Access to Information and Public Participation

Disaster risk assessments provide public authorities and other relevant actors with critical data that must be properly collected, analysed, and utilised for effective disaster prevention.

A key requirement frequently established in international instruments is the obligation to share this information with all the potentially affected individuals and communities. The public must be granted adequate access to information concerning risks. For example,

[46] Directive 2007/60/EC of the European Parliament and of the Council of 23 October 2007 on the assessment and management of flood risks OJ L 288. See also Billing's chapter in the present volume.

[47] Sendai Framework, para 24.

[48] See WMO, Early Warnings for All, available at https://earlywarningsforall.org/site/early-warnings-all#:~:text=The%20Early%20Warnings%20for%20All

the Convention on the Transboundary Effects of Industrial Accidents, adopted in Helsinki on 18 March 1992 under the auspices of the UN Economic Commission for Europe (UNECE), mandates in Article 9 that States ensure ‘adequate information is given to the public in areas potentially affected by an industrial accident arising from hazardous activities’.[49] The Convention goes further by requiring public participation in relevant procedures, enabling affected or potentially affected individuals to express their views and concerns regarding prevention and preparedness measures. Additionally, State Parties must guarantee access to justice for natural or legal persons impacted or at risk, extending such access equally to individuals residing in third States.

This right to information and participation is also enshrined in the Seveso III Directive, particularly in Articles 14 and 15.[50] Similarly, the ILO Convention 174 obliges employers to consult workers and their representatives at major hazard installations to ensure safe working systems. Similar provisions also appear in environmental agreements, including the Paris Agreement.[51]

Procedural rights relating to access to information, public participation, and access to justice are further consolidated in international human rights instruments such as the Aarhus Convention and the Escazú Agreement, supported by extensive jurisprudence.[52]

The Sendai Framework, which dedicates a specific section to stakeholder roles, including civil society, further urges States to strengthen public education and awareness through campaigns, social media, and community mobilisation, with special attention to the needs of specific audiences. In addition, the UNDRR has supported global and regional public awareness initiatives, such as the ‘One Million Safe Schools and Hospitals’ initiative and the ‘Making Cities Resilient: My City is Getting Ready!’ campaign.[53]

---

[49] Convention on the Transboundary Effects of Industrial Accidents, 17 March 1992 (UNECE Convention).

[50] Directive 2012/18/EU of the European Parliament and of the Council of 4 July 2012 on the control of major-accident hazards involving dangerous substances, amending and subsequently repealing Council Directive 96/82/EC, OJ L 197. See also Billing’s chapter in the present volume.

[51] Paris Agreement, Art 7(5).

[52] Aarhus Convention on Access to Information, Public Participation in Decision-Making and Access to Justice in Environmental Matters, 25 June 1998 and Regional Agreement on Access to Information, Public Participation and Justice in Environmental Matters in Latin America and the Caribbean, 4 March 2018. Also see the cited ECtHR jurisprudence.

[53] See UNDRR, ‘One million safe schools and hospitals campaign: advocacy guide’, available at https://www.preventionweb.net/files/15656_1msshadvguideprefinal0318101.pdf and ‘Making cities resilient: my city is getting ready!—campaign kit’, available at https://www.preventionweb.net/files/14043_campaignkit1.pdf

5.2.4 Cross-Border Information Exchange

Relevant information must also be shared at an international level, among States and with appropriate international bodies, to support coordinated efforts in disaster prevention.

An illustrative example in this regard is the Tampere Convention on the Provision of Telecommunication Resources for Disaster Mitigation and Relief Operations. This treaty regulates international cooperation in telecommunications for disaster relief, extending cooperation to the prevention phase by enabling States to share information to predict, monitor, and mitigate disasters.[54]

Furthermore, international instruments typically establish institutional bodies that regularly convene to monitor implementation and discuss evolving risks. For instance, the aforementioned UNECE Convention holds annual COPs that address emerging challenges, such as risks linked to the energy transition. Other examples include the International Atomic Energy Agency (IAEA), which oversees peer reviews and reporting under the Convention on Nuclear Safety and the Joint Convention on the Safety of Spent Fuel and Radioactive Waste Management.[55] Similarly, the International Convention on Oil Pollution Preparedness, Response and Co-operation (OPRC Convention) designates the International Maritime Organization (IMO) to collect, collate, and disseminate information from Parties, facilitating information exchange and enhancing prevention and preparedness.

Regional organisations also play a vital role in facilitating information exchange among their Member States. Beyond the EU, the Council of Europe also fosters cooperation in DRR through the EUR-OPA Major Hazards Agreement, established in 1987. This platform promotes collaboration on knowledge-sharing on prevention, preparedness, risk management, and post-crisis analysis via meetings at political and scientific levels.[56]

5.3 Preparedness in International Instruments

As noted in the previous section, prevention activities are embedded in a wide range of international instruments, particularly when taking into account 'deep prevention', that is, efforts addressing the underlying drivers and root causes of disaster risk.[57] By contrast, preparedness activities tend to be more specific and operational in nature. As a result,

---

[54] Tampere Convention on the Provision of Telecommunication Resources for Disaster Mitigation and Relief Operations, 18 June 1998. See also Wolfrum (2010).

[55] Joint Convention on the Safety of Spent Fuel Management and on the Safety of Radioactive Waste Management, 5 September 1997.

[56] EUR-OPA Major Hazards Agreement, adopted through Resolution (87) 2 of the Committee of Ministers of the Council of Europe on 20 March 1987. It has 22 Member States, most in South-Eastern Europe in addition to Portugal, as well as Morocco and Lebanon (which are not members of the Council of Europe).

[57] See Viñuales et al. (2021) for the use of such a term in the specific context of pandemic prevention.

international legal sources in this area are more limited in scope, while provisions are often more targeted.

At the global level, the Sendai Framework once again provides the most comprehensive guidance. Preparedness is specifically addressed under Priority 4: 'Enhancing disaster preparedness for effective response and to "Build Back Better" in recovery, rehabilitation and reconstruction'. This includes promoting coordinated response planning, strengthening capacities for effective emergency response, and integrating preparedness into development and recovery planning.

Some universal multilateral treaties also contain provisions relevant to preparedness. For instance, the UNECE Convention requires States to prepare and maintain emergency plans and cooperate across borders.[58] The IHR obliges State Parties to prepare for health emergencies by developing, strengthening, and maintaining the core public health capacities needed for detection and response.[59]

However, more detailed and structured commitments to preparedness are typically found in regional frameworks.[60] This is unsurprising: many regional cooperation mechanisms are specifically designed to enhance collective preparedness for shared or transboundary disaster risks. Compared to global treaties, these regional instruments often contain more specific provisions, reflecting the closer political, geographic, or institutional ties among participating States.

The clearest example of this trend is the EU, which has developed the most advanced regional disaster management system. Notably, the UCPM and the establishment of the Emergency Response Coordination Centre (ERCC) have significantly strengthened preparedness at EU level.[61]

Developments, however, are also taking place outside the 'old continent'. In Southeast Asia, the ASEAN Agreement on Disaster Management and Emergency Response (AADMER), adopted in 2005, reflects a strong regional commitment to early warning and preparedness.[62] The Agreement established the ASEAN Coordinating Centre for Humanitarian Assistance (AHA Centre), which plays a key role in operational coordination and capacity-building. In the Caribbean, the Agreement Establishing the Caribbean Disaster Emergency Management Agency (CDEMA) similarly adopts a comprehensive approach,

58 UNECE Convention, Art 8.

59 IHR, Art 13.

60 On regionalism in international disaster law, see Natoli (2021) and Shucksmith-Wesley and White (2023).

61 See the UCPM Decision, then integrated by the implementing Decision 2014/762/EU, 16 October 2014, which has been repealed in April 2025 and replaced with Commission Implementing Decision (EU) 2025/704 of 10 April 2025 laying down rules for the implementation of Decision No 1313/2013/EU of the European Parliament and of the Council on a Union Civil Protection Mechanism and repealing Commission Implementing Decisions 2014/762/EU and (EU) 2019/1310, OJ L 2025/704. These developments are discussed in greater detail in Billing's chapter in the present volume.

62 ASEAN Agreement on Disaster Management and Emergency Response, 27 July 2005. It was ratified by all the 10 Member States of the Association of Southeast Asian Nations (ASEAN).

integrating DRR policies with mechanisms for enhancing regional preparedness.[63] In Africa, while no binding regional or sub-regional treaty is exclusively dedicated to disaster preparedness, a number of regional organisations, including the African Union, have adopted relevant policies, frameworks, and soft law instruments to support preparedness within their respective mandates.[64] Moreover, the 2009 African Union Convention for the Protection and Assistance of Internally Displaced Persons in Africa (Kampala Convention) enshrines in Article 4 the general obligation to establish and implement DRR strategies as well as emergency and disaster preparedness and management measures, with a view to preventing internal displacement.[65]

Bilateral agreements also play an important role in strengthening preparedness, especially in contexts involving cross-border risks. These often include provisions on joint contingency planning, early warning systems, and mutual assistance. Moreover, treaties focused primarily on post-disaster response may also contribute to preparedness, insofar as they require States to plan for cooperation in advance.[66]

Preparedness encompasses a wide range of activities and arrangements. The following sections highlight those we consider most significant in terms of international practice and legal development: (i) developing contingency plans; (ii) conducting training activities; (iii) providing notification and warnings in case of the occurrence of accidents and other hazardous events.

5.3.1 Contingency Plans

Several multilateral conventions include the obligation to have contingency plans in place. Conventions dealing with technological hazards require the preparation and testing of both on-site and off-site emergency plans.[67]

Some Conventions further specify the characteristics of these plans. A typical feature of the plans is that they must be reviewed and updated regularly or when special circumstances arise.[68] In addition, they should set out the roles and responsibilities of organisations, describe the response actions to be taken, and identify the resources necessary to implement the plan. For example, the OPRC Convention sets out a national

[63] Agreement Establishing the Caribbean Disaster Emergency Management Agency (CDEMA), 2 July 2008.

[64] See, for example, the latest launch of the African Disaster Risk Management and Recovery Platform, available at https://www.nepad.org/news/new-era-disaster-management-launch-of-african-disaster-risk-management-and-recovery

[65] African Union Convention for the Protection and Assistance of Internally Displaced Persons in Africa (Kampala Convention), 23 October 2009.

[66] Examples of bilateral agreements can be found in the IFRC Disaster Law database, available at https://disasterlaw.ifrc.org/fr/node/219

[67] See, for example, Convention on Nuclear Safety, Art 16.

[68] See, for example, ILO Convention 174, Art 9.

contingency plan for preparedness and response. This plan outlines the organisational relationships between the various public and private bodies involved and requires oil pollution emergency plans from ships, offshore units, and the authorities responsible for seaports and other oil-handling facilities.[69] The ILO Convention 170 enshrines for employers the obligation to formulate, implement, and periodically review a documented system of major hazard control, including risk assessment, technical and organisational measures to prevent accidents, emergency plans and procedures, and schemes of consultations with workers representatives. A 'safety report' must also be prepared based on these requirements, while a designated national competent authority shall perform inspections through properly qualified and trained staff.[70]

The Sendai Framework includes the preparation of contingency plans among the first recommendations under Priority 4. It specifies that these plans must be prepared and updated periodically 'with the involvement of the relevant institutions, considering climate change scenarios and their impact on disaster risk, and facilitating, as appropriate, the participation of all sectors and relevant stakeholders'.[71]

Contingency plans are also envisaged within regional and sub-regional cooperation mechanisms. For example, Article 8 of the AADMER requires State Parties to develop these plans jointly or individually. As is common in regional agreements, this one goes into more detail regarding the cooperation it provides for. In fact, the same article also regulates 'standard operating procedures' to facilitate disaster relief cooperation, including the utilisation of military and civilian personnel, transportation and communication equipment, facilities, goods and services, as well as facilitating their transboundary movement. The South Asian Association for Regional Cooperation (SAARC) Agreement on Rapid Response to Natural Disasters includes the same provisions.[72]

5.3.2 Training

As outlined in the Introduction to this volume, the concept of 'capacity to cope' is a crucial component commonly featured in the disaster risk equation. It encompasses both the institutional capacity of disaster management systems and the ability of individuals, households, and society at large to prepare for and effectively deal with hazards. Strengthening this capacity requires targeted training for professionals and responders, as well as sustained efforts to educate the public and raise awareness of current and future risks.

---

[69] OPRC Convention, Art 3.

[70] ILO Convention 174, Art 20.

[71] Sendai Framework, para 33.

[72] South Asian Association for Regional Cooperation (SAARC) Agreement on Rapid Response to Natural Disasters, 11 November 2011.

The 2024 IFRC Disaster Risk Governance Guidelines underscore that regular training, drills, and simulation exercises are essential to ensuring disaster readiness.[73] Simulation exercises, in particular, allow responders to test their roles, responsibilities, and coordination in conditions that closely resemble real emergency scenarios.

The obligation to train relevant personnel is embedded in various international instruments. Article 11 of the Convention on Nuclear Safety states that the Parties must take adequate steps to ensure they have a sufficient number of qualified personnel who are trained and retrained. This is also included in the emergency preparedness measures set out in Article 8 of the UNECE Convention. ILO conventions place particular emphasis on ensuring that workers receive adequate training to protect both their own safety and that of others. This is evident in ILO Convention 174, which places responsibilities on both employers operating hazardous installations and competent authorities tasked with inspection and enforcement.[74]

Some conventions also require the provision of technical cooperation to other Parties. Under Article 9 of the OPRC Convention, Parties are obliged to provide technical support, including training of personnel, to those requiring assistance.

In other cases, international organisations themselves play a central role in promoting and supporting personnel training. For example, the IMO has a mandate under Article 12 of the OPRC Convention to support capacity-building, including training. Similarly, the International Civil Defence Organization (ICDO), established in 1966,[75] aims to strengthen international cooperation among national civil protection agencies and provides technical assistance, training, and other forms of support.

Cooperation in training is also well developed in regional arrangements. The UCPM serves as a primary example,[76] though not the only one. Both the AADMER (Article 18) and the Agreement Establishing the CDEMA (Article 19) include provisions on technical cooperation in the training of personnel. Bilateral agreements are another way in which States can commit to conducting joint training and drills to enhance preparedness for possible deployment.[77]

---

[73] IFRC (2024).

[74] See Articles 9, 18, and 20.

[75] Constitution of the International Civil Defence Organization, 17 October 1966. The ICDO derives from the association *Lieux de Genève*, established in 1931, which then evolved in a Non-Governmental Organization in 1958 and finally to its current form of intergovernmental organisation with the entry into force of its Constitution. At the time of writing, the ICDO comprises 60 Member States, which include mainly developing countries spanning from Southeastern Europe to Asia, Africa, and Latin America. The ICDO consists of three bodies, the General Assembly, the Executive Council, and the Permanent Secretariat.

[76] See, in particular, the Commission Implementing Decision (EU) 2025/704 of 10 April 2025, Chap. 7.

[77] For example, see Agreement between Spain and Argentina on Cooperation on Disaster Preparedness and Prevention, and Mutual Assistance in the Event of Disasters, 3 June 1988, Art V and Protocol of Intentions between the Government of the United States of America and the Republic

Complementary to professional training is the need to raise awareness and provide education on disaster risks across society. Several soft law instruments promote this broader approach. The IFRC and UNDP Checklist on Law and DRR[78] and the Sendai Framework both recommend the inclusion of disaster preparedness training in school curricula and extracurricular activities.

5.3.3 Notification and Warnings

A final fundamental preparedness activity concerns the obligation to alert all interested and potentially affected stakeholders of an imminent or recent event that may trigger a disaster.

Several international conventions dealing with technological and environmental hazards impose a duty to notify relevant parties of hazardous events that could cause a disaster, particularly those with potential transboundary impacts. This obligation is essential for enabling timely and coordinated responses across borders.

The principal example is the Convention on Early Notification of a Nuclear Accident, adopted in 1986 in the wake of the Chernobyl nuclear disaster.[79] The Convention establishes an international notification system for nuclear accidents that may give rise to transboundary radiological consequences. It requires States to promptly report key details of such accidents—including the time, location, nature, and any other relevant data needed to assess the situation—to affected States directly or via the IAEA, and to the IAEA itself. The obligation applies to accidents involving any of the facilities and activities listed in Article 1, such as nuclear reactors, nuclear fuel cycle facilities, and radioactive waste management installations. The Convention has achieved broad adherence, with 134 State Parties at the time of writing.

The International Convention for the Prevention of Pollution from Ships (MARPOL) establishes a duty to report incidental discharges involving harmful substances. This duty is elaborated through a dedicated Protocol and several annexes containing specific regulations to prevent pollution from oil, noxious liquids, sewage, garbage, and other substances.[80] Similarly, the United Nations Convention on the Law of the Sea (UNCLOS), in Article 198, also sets out a duty for States to notify other States and competent international organisations when they become aware of events or the imminent risk of marine pollution.[81] The Convention on the Law of the Non-navigational Uses of International Watercourses addresses emergency situations more broadly in Article 28. This

of the Philippines concerning Cooperation in Disaster Prevention and Management, 20 November 2001. For more examples, consult the IFRC Disaster Law database.

[78] IFRC and UNDP (2015).

[79] Convention on Early Notification of a Nuclear Accident, 27 October 1986.

[80] International Convention for the Prevention of Pollution from Ships, 2 September 1973.

[81] United Nations Convention on the Law of the Sea, 10 December 1982.

provision covers both natural and human-made hazards and imposes an obligation on the watercourse States experiencing an emergency to notify other potentially affected States without delay.[82]

Equally vital in disaster preparedness is ensuring timely alerts reach the population at risk. Effective early warning systems depend not only on detecting hazards but also on how warnings are formulated and communicated. Guidance such as the 2024 IFRC Guidelines highlights the importance of crafting warnings that are accessible, action-oriented, and tailored to diverse audiences. Rather than relying solely on technical data, alerts should convey likely impacts and offer clear instructions on protective measures. They must also be available in local and minority languages and use a combination of visual and textual elements to maximise comprehension. Dissemination should be handled by trusted authorities through multiple communication platforms, ranging from traditional media to digital technologies, ensuring outreach to remote or marginalised groups. Equally, systems should incorporate ways to confirm message receipt and understanding, while ongoing maintenance and adaptation are necessary to keep these systems reliable and inclusive.

5.4 Conclusions and Future Outlook

This chapter has shown that, despite the fragmentation that characterises international disaster law, obligations relating to disaster prevention do exist under international law, stemming from well-established principles and a range of international instruments.

Effective disaster prevention must be 'deep', that is, it must address the root causes of disasters. This requires engaging with a wide range of social, environmental, and economic factors: how we interact with nature, how we pursue development, and how we protect public health. A shift towards a more holistic and integrated approach to risk governance is increasingly discernible, where legal norms across various sectors—including environmental protection, climate change, global health, and human rights—are interpreted in a systemically coherent way. Such integration helps shape the scope and content of States' duties in the field of prevention. These duties include, inter alia, establishing appropriate legal and institutional frameworks, conducting risk assessments, ensuring public access to information and participation, and facilitating cross-border information exchange with other States and international bodies.

Preparedness, in parallel, plays a fundamental role in strengthening society's capacity to respond effectively to disaster events. This entails the availability of well-functioning institutions, trained personnel, informed and engaged populations, and comprehensive contingency planning. Crucially, it also involves the ability to issue timely warnings and to notify of the occurrence of events that may trigger disasters. While preparedness measures are often implemented at the local or national level, international cooperation is essential

[82] Convention on the Law of the Non-Navigational Uses of International Watercourses, 21 May 1997.

to support and enhance these efforts. The proliferation of regional disaster law arrangements underscores this trend. The EU, through the UCPM, offers the most advanced example, but similar instruments exist in other regions, such as the AADMER and the CDEMA.

Since 2015, the Sendai Framework, despite its non-binding nature, has played a key role in mobilising global attention around DRR and catalysing policy efforts in prevention and preparedness. As the Sendai's 'mandate' draws to a close in 2030, international discussions on its successor have already begun and are expected to intensify in the coming years.

Another significant development in this context is the UNGA's decision, on 4 December 2024, to initiate the negotiation of a legally binding international treaty on disaster management, to be concluded by the end of 2027.[83] Although the treaty is expected to focus primarily on disaster response, the draft negotiating text already includes provisions on reducing disaster risk through appropriate measures.[84] The extent to which prevention and preparedness will ultimately be incorporated into the treaty, as well as the level of State engagement and ratification, remain to be seen. Nevertheless, the treaty holds considerable promise: it may offer the first universal legal framework for disaster management, including prevention and preparedness, and provide a robust foundation for the post-Sendai agenda.

However, these legal and institutional advances must be weighed against the accelerating risk landscape. The Global Assessment Reports (GAR) reveal that the number of recorded disasters has increased fivefold over the past 50 years, with the trend continuing to rise.[85] In Europe, the recent European Climate Risk Assessment highlights a troubling gap between rising levels of risk and the adequacy of current preparedness measures, pointing to persistent shortcomings in policy implementation.[86] There is, therefore, a long way to go, particularly in terms of how States translate international instruments and obligations into meaningful action. The role of international law and institutions will be crucial in shaping a more disaster-resilient future. Whether they can live up to this challenge will depend not only on the ambition of future normative developments, but on the political will to implement them in practice and in solidarity.

---

[83] This is described in Sommario's chapter in the present volume.

[84] Article 9 of the set of 18 Articles on the 'Protection of Persons in the Event of Disasters' adopted by the International Law Commission in 2016 reads: 'Reduction of the risk of disasters. 1. Each State shall reduce the risk of disasters by taking appropriate measures, including through legislation and regulations, to prevent, mitigate, and prepare for disasters. 2. Disaster risk reduction measures include the conduct of risk assessments, the collection and dissemination of risk and past loss information, and the installation and operation of early warning systems'. See: UN, Yearbook of the International Law Commission, 2016, Vol II, Part Two, UN Doc A/CN.4/SER.A/2016/Add.1 (Part 2).

[85] See UNDRR (2023).

[86] European Environment Agency (2024) European Climate Risk Assessment, EEA Report 01/2024, p. 4.

Key Sentences

- **Shifting Priorities in International Disaster Law**—Disaster prevention and preparedness were long overshadowed by response, but they are now increasingly recognised as essential components of international disaster risk management.
- **Fragmentation and Evolution of the International Legal Framework**—No universal treaty comprehensively regulates disaster prevention and preparedness. Instead, obligations emerge from a fragmented yet expanding set of instruments in areas such as environmental law, climate change law, global health law, and human rights law.
- **Due Diligence Obligation to Prevent Disasters**—International law frames prevention as a duty of conduct: States must take all reasonable measures to reduce risk, while recognising that disasters cannot be entirely avoided and residual risk will persist.
- **Core Preventive Measures**—International instruments highlight four recurring requirements: robust legal and institutional frameworks, comprehensive risk assessments, access to information and public participation, and cross-border information exchange.
- **Preparedness as Operational Capacity**—Preparedness obligations are more concrete and operational in nature, focusing on contingency planning, training and awareness-raising, and timely notification and warning systems.
- **The Role of Regional Frameworks**—Regional and sub-regional organisations—most notably the EU, but also ASEAN and others—have developed the most detailed and structured commitments on preparedness.
- **The Sendai Framework as a Global Reference Point**—Despite its non-binding nature, the Sendai Framework for Disaster Risk Reduction 2015–2030 has become the leading global benchmark for prevention and preparedness, mobilising action across sectors and guiding state practice.
- **Future Outlook Towards Codification**—Ongoing negotiations at the UN level on an international disaster risk management treaty, as well as discussions on a post-2030 framework, may provide the first universal legal foundation for prevention and preparedness, though their impact will depend on political will and implementation.

References

Aronsson-Storrier M (2019) Exploring the foundations: the principles of prevention, mitigation, and preparedness in international law. In: Samuel K et al (eds) The Cambridge handbook of disaster risk reduction and international law. Cambridge University Press, pp 52–70

Bhandari P et al (2025) What is 'loss and damage' from climate change? 8 key questions, answered. World Resources Institute. https://www.wri.org/insights/loss-damage-climate-change

Casolari F (2018) Addressing discrimination in disaster scenarios. An international and EU law perspective. In: Giustiniani Z et al (eds) Routledge handbook of human rights and disasters. Routledge, pp 193–210

Cubie D (2024) Identifying principles of international disaster law. In: Aronsson-Storrier M, Breau SC (eds) Research handbook on disasters and international law, 2nd edn. Edward Elgar Publishing, pp 11–30

Duvic-Paoli LA (2018) The prevention principle in international environmental law. Cambridge University Press

Epiney A (2009) Environmental impact assessment. Max Planck encyclopedias of international law. https://opil.ouplaw.com/display/10.1093/law:epil/9780199231690/law-9780199231690-e1581?prd=EPIL

European Environment Agency (2024) European climate risk assessment. EEA report 01/2024. https://www.eea.europa.eu/en/analysis/publications/european-climate-risk-assessment

Fisher D (2007) Domestic regulation of international humanitarian relief in disasters and armed conflict: a comparative analysis. Int Rev Red Cross 89:345–372

Habibi R et al (2025) The 2024 amendments to the international health regulations: a new era for Global Health law in pandemic preparedness and response? J Law Med Ethics 53:47–50

IFRC (2017) Introduction to the guidelines for the domestic facilitation and regulation of international disaster relief and initial recovery assistance. https://disasterlaw.ifrc.org/sites/default/files/media/disaster_law/2020-09/1205600-IDRL-Guidelines-EN-LR.pdf

IFRC (2024) Disaster risk governance guidelines. In: Strengthening laws, policies and plans for comprehensive disaster risk management. https://disasterlaw.ifrc.org/DRMguidelines

IFRC and UNDP (2015) The handbook on law and disaster risk reduction. https://www.ifrc.org/sites/default/files/2021-07/IFRC_Handbook_Law_DRR.pdf

IPCC (2023) Climate change 2023: synthesis report. In: Lee H, Romero J (eds) Contribution of working groups I, II and III to the sixth assessment report of the intergovernmental panel on climate change. https://www.ipcc.ch/report/ar6/syr/downloads/report/IPCC_AR6_SYR_FullVolume.pdf

Kälin W (2012) The human rights dimension of natural or human-made disasters. Ger Yearb Int Law 55:119–147

Luporini R (2025) Climate change adaptation, disaster risk reduction and human rights in international law. Springer

Mitchell AD, Munro J (2023) An international law principle of non-regression from environmental protections. Int Comp Law Q 72:35–71

Natoli T (2019) Literature aligning climate change adaptation (CCA) and disaster risk reduction (DRR), IFRC|UCC. https://www.ifrc.org/document/literature-review-aligning-climate-change-adaptation-cca-and-disaster-risk-reduction-drr

Natoli T (2021) Unus pro omnibus, Omnes pro Uno: regionalization and sub-regionalization of disaster management between sovereignty and solidarity. Int Organ Law Rev 18:216–236

Ollino A (2022) Due diligence obligations in international law. Cambridge University Press

Sadeleer N (2016) The precautionary principle and climate change. In: Faure M (ed) Elgar encyclopedia of environmental law. Edward Elgar publishing, pp 20–31
Schröder M (2014) Precautionary approach/principle. Max Planck encyclopedias of international law. https://opil.ouplaw.com/display/10.1093/law:epil/9780199231690/law-9780199231690-e1603?prd=MPIL
Shucksmith-Wesley C, White ND (2023) Regionalism and Disaster Law. Yearb Int Disaster Law 4:13–20
Sommario E, Venier S (2018) Human rights law and disaster risk reduction. Questions Int Law Zoom-in 49:29–47
Spiekermann R et al (2015) The disaster-knowledge matrix. Reframing and evaluating the knowledge challenges in disaster risk reduction. International Journal of Disaster Risk Reduction 13:96–108
Staupe-Delgado R, Rubin O (2022) Challenges associated with creeping disasters in disaster risk science and practice: considering disaster onset dynamics. Int J Disaster Risk Sci 13:1–11
Toscano-Rivalta M (2020) Disaster reduction in light of COVID-19 crisis. Questions Int Law 70:37–57
UNDRR (2023) GAR special report: measuring resilience for the sustainable development goals. https://www.undrr.org/gar/gar2023-special-report
UNDRR (2025) Global status of national disaster risk reduction strategies. Governance in action: towards resilient and risk-informed development. https://www.undrr.org/2025-global-status-national-DRR-strategies
Viñuales J et al (2021) A global pandemic treaty should aim for deep prevention. Lancet 397:1791–1792
Warwick BTC (2019) Unwinding retrogression: examining the practice of the committee on economic, social and cultural rights. Hum Rights Law Rev 19:1–24
Wolfrum R (2010) Cooperation, international law of Max Planck encyclopedias of international law. https://opil.ouplaw.com/display/10.1093/law:epil/9780199231690/law-9780199231690-e1427?prd=MPIL
Zorzi Giustiniani F (2018) Something old, something new: disaster risk reduction in international law. Questions of International Law 49:7–27
Zorzi Giustiani F et al (eds) (2018) Routledge handbook of human rights and disasters. Routledge

Open Access This chapter is licensed under the terms of the Creative Commons Attribution 4.0 International License (http://creativecommons.org/licenses/by/4.0/), which permits use, sharing, adaptation, distribution and reproduction in any medium or format, as long as you give appropriate credit to the original author(s) and the source, provide a link to the Creative Commons license and indicate if changes were made.

The images or other third party material in this chapter are included in the chapter's Creative Commons license, unless indicated otherwise in a credit line to the material. If material is not included in the chapter's Creative Commons license and your intended use is not permitted by statutory regulation or exceeds the permitted use, you will need to obtain permission directly from the copyright holder.

Conclusions and Outlook

6

Emanuele Sommario and Peter Billing

Abstract

This book examines the evolving architecture of disaster governance, focusing on the institutional, operational, and legal mechanisms that enable effective disaster management across local, national, regional, and global levels. Particular attention is devoted to the legal and institutional frameworks that regulate cooperation among states, international organisations, humanitarian actors, and local authorities throughout the disaster cycle. The analysis highlights the importance of comprehensive risk assessments, contingency planning, early-warning systems, information-sharing mechanisms, and clearly defined responsibilities among stakeholders. It also explores how international and regional legal instruments contribute to strengthening preparedness and response capacities, while promoting accountability, transparency, and coordination. Drawing on experiences from regional mechanisms such as the European Union Civil Protection Mechanism, ASEAN's Agreement on Disaster Management and Emergency Response, and the Caribbean Disaster Emergency Management Agency, the volume illustrates how legal frameworks and institutional innovation can jointly enhance operational readiness and resilience. Particular consideration is given to ongoing efforts to develop a universal legal framework for disaster management and to discussions surrounding the post-2030 successor to the Sendai Framework for Disaster Risk Reduction. The analysis concludes that effective disaster governance requires a combination of robust

E. Sommario
Scuola Superiore Sant'Anna, Pisa, Italy
e-mail: emanuele.sommario@santannapisa.it

P. Billing (✉)
Department of Geography, University of Bonn, Bonn, Germany
e-mail: pbillng@uni-bonn.de

© The Author(s) 2026

E. Sommario and P. Billing (eds.), *International Legal and Institutional Frameworks for Disaster Risk Management*, Civil Protection, Disaster Management, and Resilience 4,
https://doi.org/10.1007/978-3-032-11806-6_6

legal foundations, capable institutions, and sustained political commitment. While legal norms provide the framework for cooperation and accountability, their effectiveness ultimately depends on the capacity of institutions to coordinate resources, anticipate risks, and deliver timely and needs-based assistance to affected populations.

Keywords

Disaster governance • Disaster law • Disaster Risk Reduction (DRR) • International cooperation

The international community currently lacks a comprehensive, legally binding instrument of universal scope that encompasses all phases of the disaster cycle. Resolution 79/128, adopted by the UN General Assembly in December 2024,[1] establishes a clear mandate to negotiate such an instrument by the end of 2027. Although the negotiation process is likely to encounter contentious issues—such as the obligation to request external assistance when national capacities are exceeded—the current context presents a timely opportunity for the adoption of a treaty that would provide greater clarity regarding the roles and expectations of states and humanitarian organisations in the aftermath of disasters. The need for such a development is particularly pressing, as this volume illustrates, given the fragmented nature of the existing legal framework governing disaster management. Relevant norms are dispersed across a range of legal instruments, including those pertaining to environmental protection, public health, human rights, and disaster-specific regimes. A unified treaty would serve to consolidate these provisions, thereby clarifying the legal obligations of states, humanitarian actors, and international bodies. Such consolidation is increasingly essential in an era marked by the growing frequency and severity of disasters driven by climate change, which exacerbates both human suffering and material losses. However, the added value of any future treaty will ultimately depend on the extent of State engagement throughout the negotiation process, as well as on the breadth of ratification and subsequent implementation.

The need for a firm international legal framework also pertains to disaster prevention and preparedness. This framework must address the root causes of disaster to be effective. It requires tackling a wide range of social, environmental, and economic factors drivers of risk. The Sendai Framework—albeit non-binding in nature—has played a key role in mobilising global attention around DRR and catalysing policy efforts in prevention and preparedness. As its 'mandate' draws to a close in 2030, international discussions on its successor have already begun and are expected to intensify in the coming years. The draft negotiating text of the binding legal instrument mentioned above already includes provisions on reducing disaster risk through appropriate measures. Preparedness plays a crucial role in enhancing a society's capacity to respond effectively when disasters strike.

[1] UNGA, Protection of persons in the event of disasters, UN Doc A/RES/79/128, 12 December 2024.

It relies on well-functioning institutions, trained personnel, informed communities, robust contingency planning, and tailored early warning systems.

Although the extent to which prevention and preparedness will ultimately be incorporated into the treaty remains to be seen, it may offer the first universal legal framework for disaster management, including prevention and preparedness, and thus provide a robust foundation for the post-Sendai agenda. Such cross-sectoral integration helps shape the scope and content of states' duties in the field of prevention and preparedness. Key requirements should include the following:

- Establishing appropriate legal and institutional frameworks for risk reduction and resilience.
- Conducting comprehensive multi-hazard risk assessments.
- Ensuring public access to information and participation opportunities.
- Facilitating cross-border information exchange and international cooperation.

International cooperation remains indispensable to strengthening national and local preparedness efforts. Regional mechanisms such as the European Union's Civil Protection Mechanism (UCPM), ASEAN's AADMER, or the Caribbean's CDEMA demonstrate the added value of tailored, cooperative frameworks. The proliferation of regional disaster law arrangements underscores this trend. The EU's UCPM currently offers the most advanced example in that respect. It illustrates how iterative legal and financial adaptations can yield a versatile, responsive disaster mechanism. Its 2024 evaluation highlights[2] significant achievements, such as enhanced coordination (ERCC), rapid response capacities (rescEU, ECPP), and robust expert networks. It also identifies areas for further improvement, such as simplifying administrative procedures and shoring up human and financial resources. The evaluation also acknowledges that important adaptations and promising initiatives can be expected to yield future benefits, such as implementation of the Disaster Resilience Goals, the creation of additional reserve capacities, as well as ongoing work on anticipation, analysis and scenario-building. In light of geopolitical developments, civil defence aspects are likely to gain in importance. These issues are addressed in the Commission proposal of 16 July 2025 for a new Regulation on the Union Civil Protection Mechanism aiming—inter alia—at enhancing coordination across sectors, including civil-military cooperation.[3]

---

[2] European Commission, Communication from the Commission to the Council and the European Parliament on the evaluation of the Union Civil Protection Mechanism-Strengthening EU's emergency preparedness, COM (2024) 212 final, 29 May 2024.

[3] European Commission, Proposal for a Regulation of the European Parliament and of the Council on the Union Civil Protection Mechanism and Union support for health emergency preparedness and response, and repealing Decision No 1313/2013/EU (Union Civil Protection Mechanism), COM (2025) 548 final, 16 July 2025.

At the global level, emergency and crisis management are becoming increasingly complex, with a growing number of actors competing for resources and influence. Ensuring coherence between strategic and operational tiers, and across global, regional, national, and local levels, is the fundamental challenge. Effective coordination must become part of the solution and minimise the risk of overlap or duplication. Maintaining clear accountability to affected populations, domestic and international stakeholders demands agreed standards, transparent evaluation indicators, and rigorous needs assessments. In complex emergencies, civil-military coordination must adhere to the principles of *last resort* and, more generally, *do no harm*, reinforcing humanitarian norms and excluding actors unwilling or unable to respect them.

In terms of outlook, over the next 3 years, states will negotiate the first universal treaty on disaster management. The core issues will include the balance between State sovereignty and mandatory assistance obligations, the treaty's scope across prevention, preparedness, and response, and possible mechanisms for enforcement and dispute resolution. In addition, as the Sendai Framework era ends in 2030, its successor will need to address gaps in risk governance left by a non-binding mandate. Embedding legally binding prevention and preparedness obligations into the new global framework could align incentives and resources towards long-term resilience.

The current disaster trends underscore the urgency of translating normative advances into practice. Political will at all levels will determine whether international commitments are mirrored by national legislation, institutional reforms, and investments in risk reduction. Key actions for the coming decade should thus include the following:

- Finalising the UN-led disaster management treaty by 2027.
- Developing a robust post-Sendai framework with binding targets for prevention and preparedness.
- Harmonising international, regional, and national legal instruments to eliminate regulatory gaps.
- Strengthening monitoring and evaluation systems with clear, verifiable indicators.
- Promoting community engagement in line with Grand Bargain commitments to ensure needs-driven responses.

Success will depend not only on the ambition of legal instruments but on genuine solidarity among states, international organisations, and civil society. While important challenges remain, the next decade offers a unique opportunity to transform fragmented norms into an integrated, action-oriented architecture capable of reducing disaster risk and delivering timely relief to vulnerable populations.

Open Access This chapter is licensed under the terms of the Creative Commons Attribution 4.0 International License (http://creativecommons.org/licenses/by/4.0/), which permits use, sharing, adaptation, distribution and reproduction in any medium or format, as long as you give appropriate credit to the original author(s) and the source, provide a link to the Creative Commons license and indicate if changes were made.

The images or other third party material in this chapter are included in the chapter's Creative Commons license, unless indicated otherwise in a credit line to the material. If material is not included in the chapter's Creative Commons license and your intended use is not permitted by statutory regulation or exceeds the permitted use, you will need to obtain permission directly from the copyright holder.

The manufacturer's authorised representative in the EU is Springer Nature Customer Service Centre GmbH, Europaplatz 3, 69115 Heidelberg, Germany. If you have any concerns regarding our products, please contact ProductSafety@springernature.com

Printed and bound by CPI Group (UK) Ltd, Croydon, CR0 4YY

05/10/2026

02228041-0001